SWIRLING FLOW PROBLEMS AT INTAKES

INTERNATIONAL ASSOCIATION FOR HYDRAULIC RESEARCH
ASSOCIATION INTERNATIONALE DE RECHERCHES HYDRAULIQUES

IAHR AIRH

1

HYDRAULIC STRUCTURES DESIGN MANUAL

HYDRAULIC DESIGN CONSIDERATIONS

SWIRLING FLOW PROBLEMS AT INTAKES

Coordinator-editor
JOST KNAUSS
Technical University of Munich, Obernach, FR Germany

CRC Press
Taylor & Francis Group
Boca Raton London New York

CRC Press is an imprint of the
Taylor & Francis Group, an **informa** business
A TAYLOR & FRANCIS BOOK

Cover photograph: From *Das sensible Chaos. Strömendes Formenschaften in Wasser und Luft,* by Theodor Schwenk; with permission of Verlag Freies Geistesleben G.m.b.H., Stuttgart.

First published 1987 by Taylor & Francis

Published 2007 by CRC Press
Taylor & Francis Group
6000 Broken Sound Parkway NW, Suite 300
Boca Raton, FL 33487-2742

ISBN-13: 978-90-6191-643-7 (hbk)

Visit the Taylor & Francis Web site at
http://www.taylorandfrancis.com

and the CRC Press Web site at
http://www.crcpress.com

Publisher's Note
The publisher has gone to great lengths to ensure the quality of this reprint
but points out that some imperfections in the original may be apparent

Contents

Preface

This volume is the first of a series. It belongs to the IAHR *Hydraulic structures design manual* which is being prepared by a team of authors co-ordinated by the IAHR, as was decided during the IAHR Congress in New Delhi in 1981. This manual is to appear in the form of two series. The first series will present up-to-date information, worldwide, on the basic knowledge and tools available for the analysis of phenomena affecting hydraulic structures. The second series will serve the practising engineer as an application index aiding him to select the relevant sections of the manual to be consulted in the task to ensure a hydraulically safe design:

Hydraulic design considerations:
Discharge characteristics, energy losses
Hydrodynamic forces
Flow-induced vibrations
Cavitation
Air entrainment in free-surface flows
Air water mixture problems in closed conduits
Swirling flow problems at intakes
Energy dissipation
Scouring
Sediment control
Seepage and uplift
Water-equality effects.

Application index:
Low-head control structures
High-head control structures
Intake structures
Conduit and channel transitions
Energy dissipators
Fish protection facilities.

On behalf of the Editorial Committee, consisting of J.S.McNown, J.E.Prins, A.J.Raudkivi, D.Vischer, and myself, I am happy to present this first contribution to the IAHR Design manual for the benefit of all those who are involved with hydraulic structures both in engineering practice and education. Work on the remaining parts of the manual is well on its way. The other volumes will come out within the next two to three years as they are completed. I am sure the engineering community will welcome the consolidation of this information which is normally scattered among a vast number of journals and books, with the assurance that both the data collection and its presentation were performed on an international, highly-expert level.

The present volume *Swirling flow problems at intakes* is a good example. It was prepared by eleven authors from different countries who are all internationally recognized experts on swirling flow problems. Moreover, Dr. Knauss, the coordinator-editor, has taken great pains to receive input to this work from discussions at international meetings. When the manuscript was completed, he had it reviewed by five experts in the field. Despite all these efforts, I wish to second the words of caution expressed by Dr. Knauss and Dr. Hecker in the Introduction and the Conclusions. There is still intensive research activity today on swirling flows so that it is impossible to present material in a final form. Therefore, rather than feigning false security by stringent guidelines and prescribed methods, the monograph presents 'state-of-the-art' knowledge by means of a number of summarizing contributions. We wish to invite all colleagues working in this field to report to us new findings, from both research and practical experience, so that we may incorporate them in the revised new editions which are planned to appear in five year-intervals.

Last, but not least, I wish to thank everyone who helped to make publication of this volume possible. In particular, I wish to thank Dr. Knauss who had the tremendous and often frustrating task of coordinating and editing the many contributions. We are most grateful to him and all his co-authors for taking on the chore of presenting their knowledge and experience, and for doing it in such an admirable form, even though their jobs already occupied them to the limit. The secretariat of IAHR is to be thanked for giving the much-needed support and counsel in all phases of the publication of this volume.

Karlsruhe, May 1987 Eduard Naudascher

Contributors

E.Chang, Dr.
Training Centre for Small Hydro Power
P.O.Box 544
Hangzhou
China

R.Ettema, Assi. Prof. Dr.
Iowa Institute of Hydraulic Research
University of Iowa
Iowa City, Iowa 52242
USA

R.J.Garde, Prof. Dr.
Department of Civil Engineering
University of Roorkee
Roorkee, U.P.
India

G.E.Hecker, Prof. Dir.
Alden Research Laboratory
Worcester Polytechnic Institute
Holden, Massachusetts 01520
USA

S.C.Jain, Prof. Dr.
Iowa Institute of Hydraulic Research
University of Iowa
Iowa City, Iowa 52242
USA

J.Knauss, Prof. Dr.-ing.
Versuchsanstalt für Wasserbau
Technische Universität München
D-8111 Obernach/Walchensee
FR Germany

M.Padmanabhan, Dr.
Alden Research Laboratory
Worcester Polytechnic Institute
Holden, Massachusetts 01520
USA

M.J.Prosser
Principal Research Engineer
BHRA Fluid Engineering
Cranfield, Bedford
UK

K.G.Ranga Raju, Prof. Dr.
Department of Civil Engineering
University of Roorkee
Roorkee, U.P.
India

P.Rutschmann
Versuchsanstalt für Wasserbau
ETH-Zentrum
CH-8092 Zürich
Switzerland

D.Vischer, Prof. Dr.
Versuchsanstalt für Wasserbau
ETH-Zentrum
CH-8092 Zürich
Switzerland

P.Volkart, Dr.
Versuchsantstalt für Wasserbau
ETH-Zentrum
CH-8092 Zürich
Switzerland

Reviewers

F.W.Blaisdell
Research leader
Agricultural Research Service
St. Anthony Falls Hydraulic Laboratory
Minneapolis, Minnesota 55414
USA

G.S.Dhillon, Dr.
Superintending engineer
Anandapur Sahib Hydel Const., Circle No. 2
Nangal Township, Pb.
India

Santokh Singh
Chief engineer (Research)
Irrigation and Power Research Institute
Amritsar, Pb.
India

Symbols and dimensionless numbers

Symbols used in this volume correspond to ISO 772. Symbols which are not mentioned below will be introduced within the individual chapters in which they are first mentioned.

$g\ (m \cdot s^{-2})$	acceleration due to gravity
$\rho\ (kg \cdot m^{-3})$	fluid mass density
$\nu\ (m^2 \cdot s^{-1})$	kinematic viscosity
$\sigma\ (N \cdot m^{-1})$	surface tension force per unit length
$h\ (m)$	intake submergence (detailed definition given in Figure 4.1)
$h_{cr}\ (m)$	critical intake submergence
$d\ (m)$	intake diameter (or an other typical linear dimension of the intake cross section if not circular)
$Q\ (m^3 \cdot s^{-1})$	intake discharge
$v\ (m \cdot s^{-1})$	intake velocity: $v = \dfrac{Q}{d^2\pi/4}$
$v_t\ (m \cdot s^{-1})$	tangential velocity: $v_t = \omega\, r$, with ω = angular velocity, r = general radius, and d/2 = intake radius, r_1 = matching radius between free and forced vortex region (see Figures 2.1 and 4.8), r_o = shadow radius (outer radius of surface depression, see Figure 5.4)
$\Gamma\ (m^2 \cdot s^{-1})$	circulation: $\Gamma = 2\pi\, v_t\, r = 2\pi\, c$
$c\ (m^2 \cdot s^{-1})$	circulation constant: $c = \Gamma/2\pi$
α	swirl angle
α_a	approach flow swirl angle (index a indicates approach flow conditions)
$\bar{\varphi}$	intake orientation (detailed definition given in Figure 4.1)

Basic dimensionless numbers

h/d	relative intake submergence
$(h/d)_{cr}$	critical value
$Fr = \frac{v}{\sqrt{g\,d}}$	intake Froude number
$Re = \frac{v\,d}{\nu}$	intake Reynolds number
$We = v\sqrt{\frac{\rho\,d}{\sigma}}$	intake Weber number
$N_\Gamma = \frac{\Gamma\,d}{Q}$	circulation number

Other important notations

$C = \frac{v}{\sqrt{2g\,h}}$	discharge coefficient
$\frac{v}{\sqrt{g\,h}} = \frac{1}{\sqrt{h/d}}\,Fr$	submergence Froude number
$\frac{Q}{\nu\,h} = \frac{4}{\pi}\,\frac{1}{h/d}\,Re$	radial Reynolds number
$\frac{v^2\,\rho\,d}{\sigma} = We^2$	Weber number

Other notations of circulation number

$\frac{\Gamma}{v\,d} = \frac{\pi}{4}\,N_\Gamma$	Kolf number
$N_c = \frac{c}{v\,d} = \frac{1}{8}\,N_\Gamma$	
$\frac{\Gamma\,h}{Q} = \frac{h}{d}\,N_\Gamma$	Jain number
$\frac{c\,d}{Q} = \frac{1}{2\pi}\,N_\Gamma$	Anwar number

Introduction 1

J.KNAUSS
Technical University of Munich, Obernach, FR Germany

1.1. SCOPE OF THE VOLUME

According to the main guide-line of the IAHR-initiative concerning the publication of a ***Hydraulic structures design manual*** this volume is intended to provide up-to-date information on international scale on the subject under discussion, namely: Swirling flow problems at intakes.

Swirling flow problems are associated with intake structures of different types which are prone to vortex formation. Problems arise from air entrainment and swirl intrusion at the transition from free surface to pressure flow conditions. Intakes confronted with such phenomena are installed at water power plants or pump sumps for example.

Generally, an intake structure may be considered as well designed when a possible vortex formation is controlled in such a way that air entrainment is avoided on principle and swirl entrainment is minimized down to a tolerable amount.

In the case of swirling flow problems at intakes the development and improvement of the basic knowledge of the hydraulic phenomena involved and of the fundamental tools available for practical application is progressing, but by no way finished. The present state of the art (1985) will be discussed in this monograph by means of a series of summarizing contributions (chapters of the monograph) presented by experienced hydraulic engineers from different countries.

1.2. SOURCES OF SWIRL

Among the many sources of swirl and vortex formation, the most frequently encountered one is an eccentricity of the approach flow relative to the intake and the connected conduit. In most of the cases, asymmetry in the approach flow field due to geometrical conditions causes vortex formation, but there are also special hydraulic processes responsible for swirl production in geometrically symmetric installations, even with an ideal velocity distribution in the approach flow. Durgin & Hecker (1978) define three fundamental types of vorticity sources

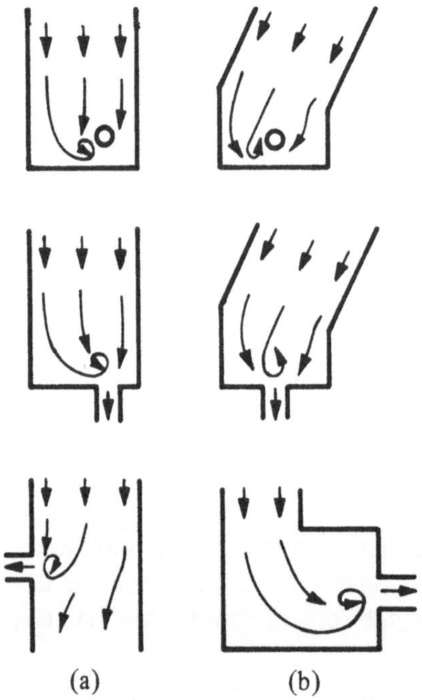

Figure 1.1. Rotational flow arising from (a) asymmetry and (b) change in direction of boundaries (Denny & Young, 1957).

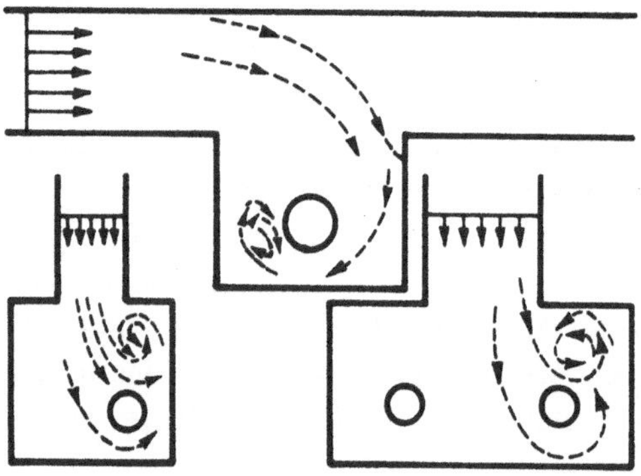

Figure 1.2. Intake prone to vortex formation (Anwar, 1968).

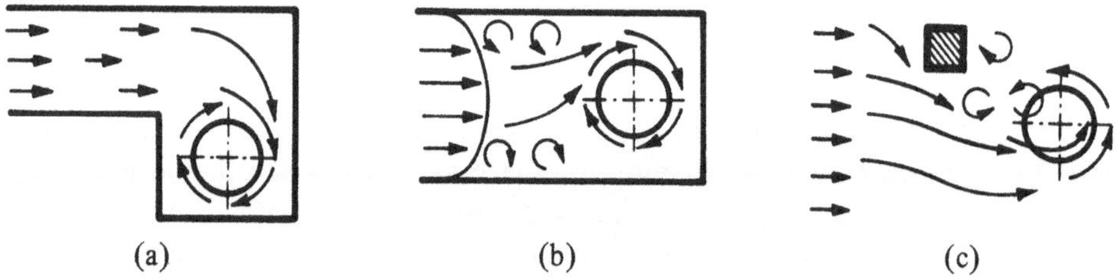

Figure 1.3. Sources of vorticity: (a) offset introduction, (b) velocity gradients, (c) obstruction (Durgin & Hecker, 1978).

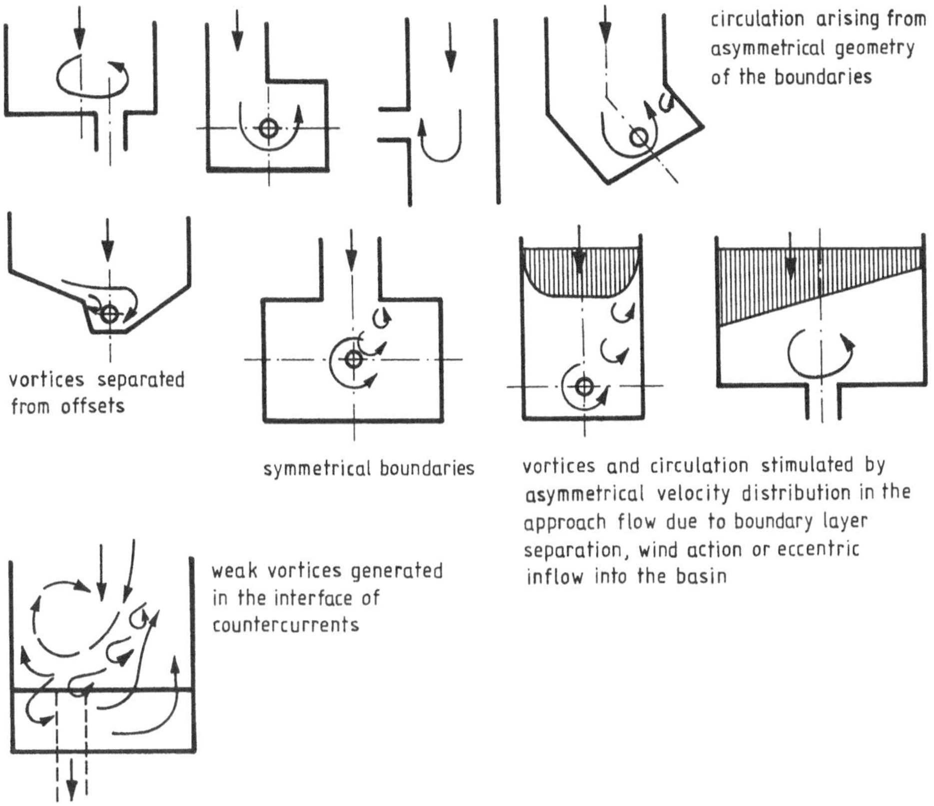

Figure 1.4. Various examples of vortex formation at asymmetrical approach flow conditions (Knauss, 1972).

as shown in Figure 1.3 and described in Section 6.2.1. In literature one can find a lot of schematic illustrations showing the various examples of intakes prone to vortex formation (see Figures 1.1 up to 1.5). Vortices mainly were stimulated by: eccentric orientation of the intake relative to a symmetric approach flow area, asymmetric approach flow conditions due to irregularities in boundary lining, unfavourable effects of obstructions such as offsets, piers or dividing walls, non-uniform velocity distribution caused by boundary layer separation, wind action, wakes, countercurrents and other reasons.

An eccentricity between the approach flow area and the intake may stimulate formation of a circulating or swirling flow field in front of or above the inlet structure. By this circulation harmless or harmful vortices may be generated. Generally one can state that the intensity and stability of the resulting flow rotation increase with the increase in area prone to circulation.

Very often the given local conditions or special design considerations such as an economic optimization of the whole scheme interfere with the hydraulic principle of a vortex free intake. Then, vortex formation must be tolerated, provided that

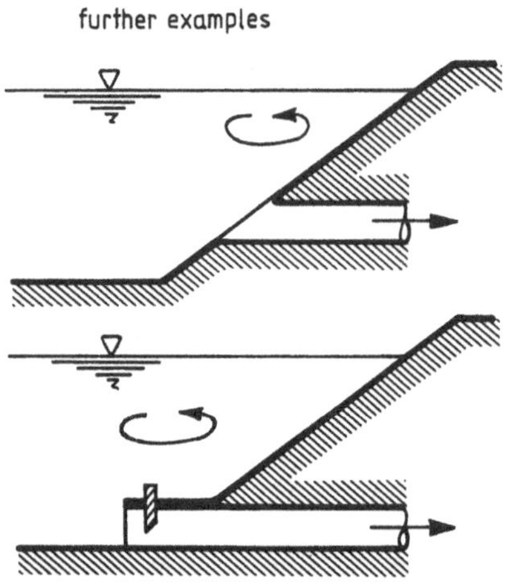

Figure 1.5. Vortex formation at intakes of pumped storage schemes (Knauss, 1983).

operational problems due to dangerous air and swirl entrainment can be eliminated. This means that circulation must be controlled and limited by suitable measures.

1.3. INTAKE TYPES, GENERAL CLASSIFICATION

The common brief classification concerns a distinction between 'intake structures' and 'pump sumps'. Intake structures are associated with power plants, especially with pumped storage schemes, with all kinds of water supply systems, with lock filling installations or with the various hydraulic structures at dams (e.g. bottom outlets, morning glory or siphon spillways). Concerning the flow rate all kinds of pump intakes in sumps may generally be considered as small or medium size installations, whereas intake structures normally are classified as large size installations.

With respect to hydraulic considerations the above mentioned classification is insufficient and should be completed by the introduction of further criteria concerning special characteristics of the construction.

Figure 1.6 shows a proposal for a subclassification due to two typical design elements. The first one is a distinction related to the intake direction or orientation. The second one considers an important structural measure distinguishing whether an intake is located in the floor or walls of the basin or is projecting into the reservoir or sump. The latter distinction may be helpful concerning the different types of vortex flows at intakes.

In Figure 1.6 also the general feature of the main hydraulic system 'intake' is pointed out schematically. Within the intake the flow conditions change from free surface to pressure flow.

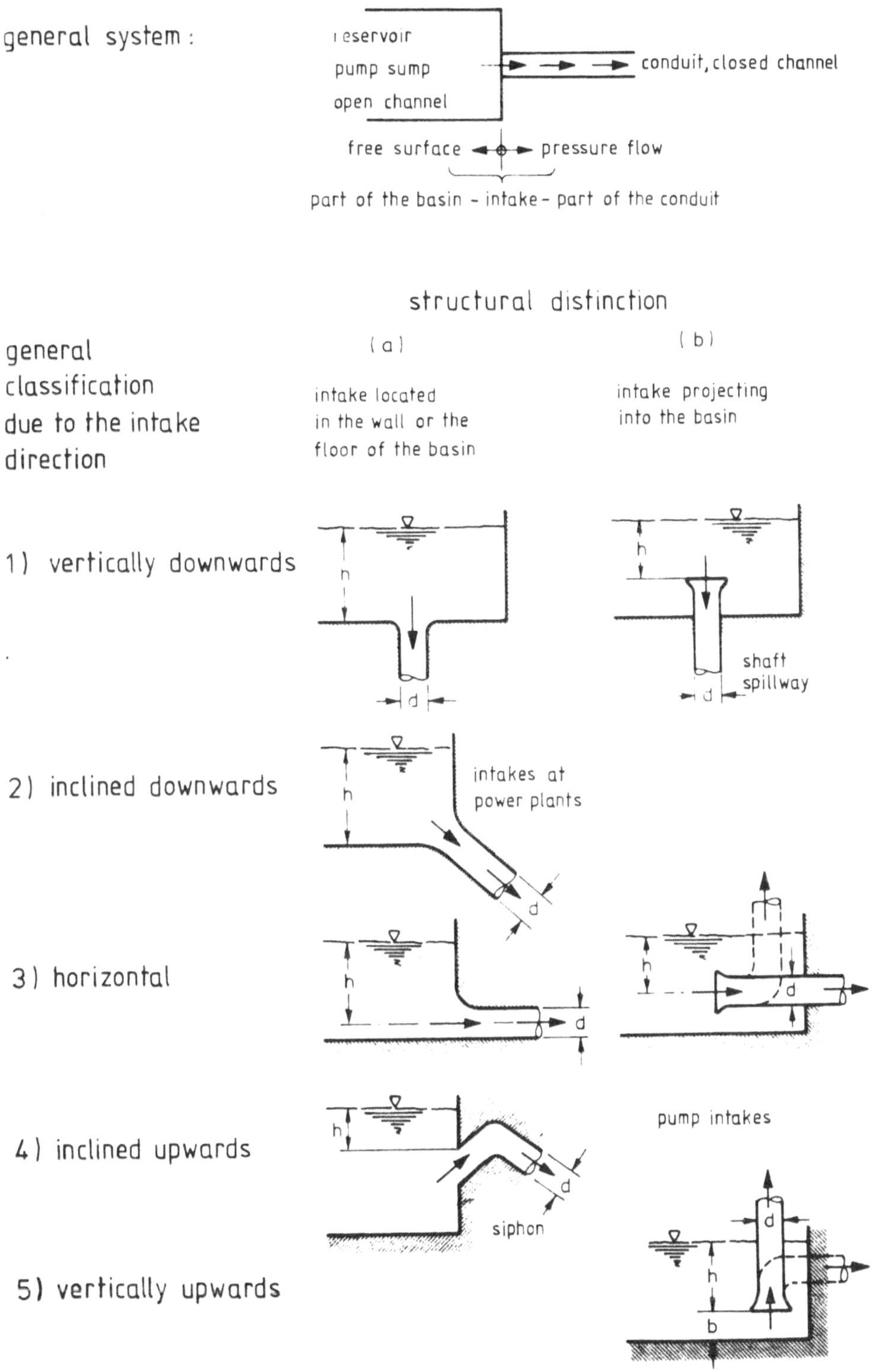

Figure 1.6. Types of intake structures.

1.4. VORTEX TYPES, GENERAL CLASSIFICATION

Vortices at intakes appear in forms corresponding to numerous type specifications and descriptions. Three main distinctive marks are given by (a) the location of vortex formation relative to the intake, (b) the time dependency of the phenomena and (c) the shape of the established vortex. Further fundamental classification is necessary concerning a vortex strength scale and an appropriate description of the vortex development relative to the intake (see Section 2.1.3).

A very important classification of vortex types considers the location of vortex formation: (a) surface vortices, starting from the free water surface and being responsible for possible entrainment of air and swirl, (b) subsurface vortices, starting from the floor and/or the walls of the upstream basin, being mainly responsible for swirl intrusion (see Figure 1.7). Wall located intakes normally are associated with surface vortices only. Projecting intakes especially in pump sumps are subjected to both, surface and subsurface vortices.

Cocerning time dependency of vortex formation, a basic distinction between steady or stable vortices and unsteady or unstable, transient, intermittent types is necessary. Stable vortices are characterized by a steady production of swirl in

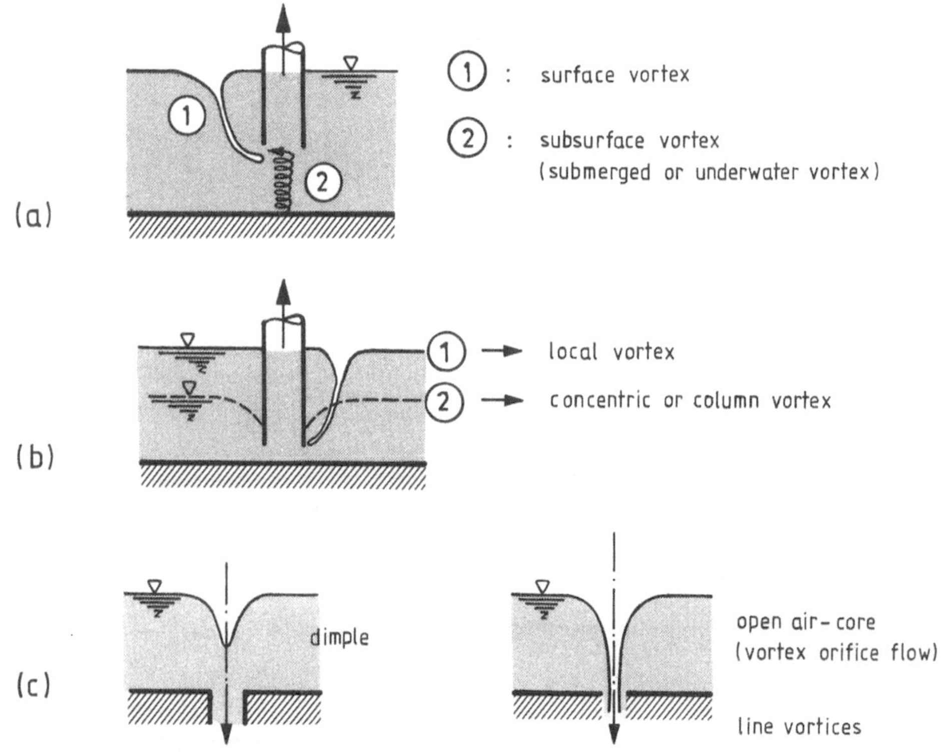

Figure 1.7. Types of vortices due to location of appearance and resulting shape.

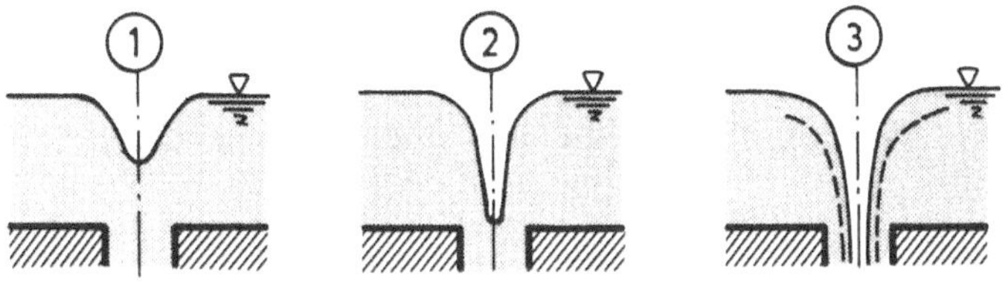

description due to:	①	②	③
1) increase of circulation	weak	critical	strong
2) development relative to the intake	undeveloped	critical	developed
3) physical explanation	combined potential and rotational vortex (Rankine vortex)	critical stage for air-entrainment	potential vortex or free spiral vortex with an open air-core

Figure 1.8. The different types of stable surface vortices at intakes vertically downwards (drain vortices), indication and distinction.

the approach flow field. Intermittent vortices result from an unsteady supply of swirl. The instability of the process may be checked by statisticaion of vortex types considers the location of vortex formation: (a) surface vortices, starting from the free water surface and being responsible for possible entrainment of air and swirl, (b) subsurface vortices, starting from the floor and/or the walls of the upstream basin, being mainly responsible for swirl intrusion (see Figure 1.7). Wall located intakes normally are associated with surface vortices only. Projecting intakes especially in pump sumps are subjected to both, surface and subsurface vortices.

Cocerning time dependency of vortex formation, a basic distinction between steady or stable vortices and unsteady or unstable, transient, intermittent types is necessary. Stable vortices are characterized by a steady production of swirl in the approach flow field. Intermittent vortices result from an unsteady supply of swirl. The instability of the process may be checked by statistical methods as discussed in Section 2.1.4.

Some common type classifications of stable surface vortices due to the overall shape of the established vortex and its development relative to the intake are shown in Figures 1.7 and 1.8. The Rankine combined vortex is discussed in Sections 2.1.2 and 4.4.1. Information on free spiral vortices is given in Chapter 3.

1.5. SWIRLING FLOW PROBLEMS

Two main hydraulic problems may be encountered with the entrainment of air and swirl at intakes: (a) unfavourable hydrodynamic impact on the operation an performance of hydraulic machines, (b) dangerous hydropneumatic effects in the closed channel system downstream of the inlet.

An intake prone to vortex formation resulting in a considerable entrainment of air and swirl may be the source of the following difficulties (see also Sections 2.3.1 and 6.2.3):

— increase of head losses due to the hindered intake process,

— reduction of maximum intake rate by the vortex flow itself or by means of adjusted operational conditions aimed to minimize swirling flow problems,

— reduction of efficiency of the hydraulic machines due to the decreased discharge and by operating the machines within an unfavourable region of efficiency occurring at lower flow rates,

— deterioration of the running conditions of the hydraulic machines by entrained air associated with fluctuations of flow rate and pressure distribution, at least resulting in considerable drops of efficiency, rough operation conditions and environmental impact by increased noise,

— stimulation of vibrations and cavitation inception leadin to wear and damage of sensitive construction elements (impeller blades, bearings or other important parts of the installation),

— production of an unbalanced loading of pump impellers or turbine runners by swirl and uneven flow distribution,

— appearance of slug flow conditions within the conduit due to accumulation of entrapped air (sudden release of large air pockets, flow pulsations),

— increase in operational problems by the suction of debris possibly pulled down by surface vortices (e.g. asymmetric blockage of screens, dislocation of bulk head and damage to grating).

All the listed difficulties may be intensified by the normally given instability of the vortex forming process. The most disagreeable consequences are associated with the entrained air.

Ideally, the flow passing from free surface to closed channel conditions should be uniform, steady and of single phase (Prosser, 1977). The latter condition concerns two main postulations for intake design and operation: no direct entrainment of free air by surface vortices and no release of dissolved air or gas by strong subsurface vortices. However, for practical application the desirable uniformity and absolute steadyness of the intake flow pattern must be modified by the introduction of tolerable limits and acceptance criteria handling unavoidable swirl entrainment.

1.6. QUANTIFICATION OF DANGEROUS AIR AND SWIRL ENTRAINMENT

Generally, air entrainment causes detrimental effects on the operation of pumps and turbines, but in special cases small volumes of entrapped air may not seriously degrade the performance conditions (see Section 2.2.3). Axial flow pumps are more sensitive to air ingestion than centrifugal pumps (Chang, 1977). Air by water volumes up to 3 or 4% may give rise to a small but continuous decrease in pump efficiency.

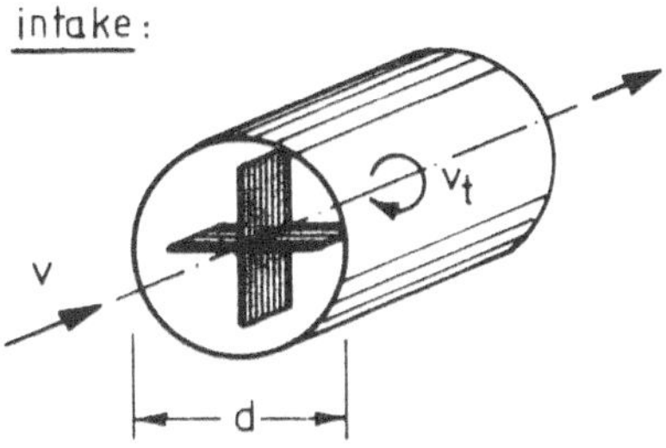

n = revolutions of swirl meter per time

$$v_t = \pi d n, \quad v = \frac{4\,Q}{d^2 \pi}$$

$$\mathrm{tg}\,\alpha = \frac{v_t}{v}, \quad \alpha^{\circ} = \text{indicated swirl angle}$$

Figure 1.9. Swirl meter (see Section 2.2.2).

Ingestion rates above this margin normally result in a sudden drop of efficiency. For an interruption of pump operation large air rates of about 7 up to 20% are necessary. Critical operation conditions vary with the pump type and the characteristic performance criteria. Fundamentally, for a 1% drop in efficiency only a small rate of entrapped air is necessary. But a loss in efficiency by this relatively small rate may lead to losses in profit which, in a few years, can exceed the initial capital costs of the pump (Chang, 1979b).

An interesting investigation on the influence of vortex entrained air on the operational conditions of a turbine (Dexter & Zeigler, 1978) did not result in untenable pressure surges or runner deflections. However, interrupted air ingestion did create rough operational conditions, which can be tolerated only over a limited time.

In summary, vortex entrained air normally disturbs the safe operation of hydraulic machines as well as the water transport in closed channel supply systems. Air entrainment by intake vortices should be avoided on principle.

In order to define and measure the entrained swirl, the swirl angle indicated by a swirl meter was found to provide a suitable criteria (see Figure 1.9 and Section 2.2.2).

Manufacturers and designers of pumps, turbines and pump-turbines postulate an upper limit of a tolerable swirl angle. Regarding the type of the hydraulic machine, prerotation expressed by the indicated swirl angle shall normally not exceed 5° (see Sections 2.3.1 and 6.2.3) and in special cases even 2° (e.g. Nathan, 1981; Maynord, 1982). Besides the observation of the tolerable prerotation affected by the swirl, uneven velocity distributions in front of the impeller must be kept within a margin of ±5% deviation from the mean value.

1.7. BASIC CONCEPT OF CRITICAL INTAKE CONDITIONS

Whether a surface vortex resulting from a stable initial circulation is air entraining or not depends very much on the water depth above the intake under consideration. Presuming a sufficient definition of the inception of air ingestion (see Section 2.1.3)

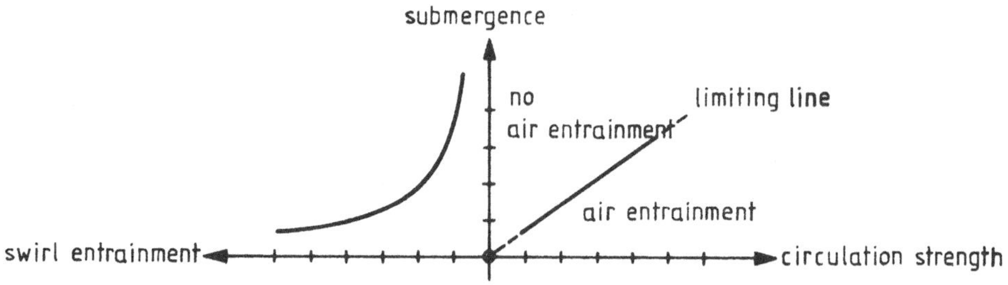

Figure 1.10. Basic relationship between circulation strength, submergence, air and swirl entrainment.

a certain critical submergence depth can be related to the given circulation strength (see Figure 1.10). The critical submergence increases with increasing circulation strength. One important design criterion concerning advoidance of air entrainment by intake vortices is to choose submergence requirements above the critical stage.

Regarding the entrained swirl, with otherwise equal conditions the indicated swirl angle is independent of the flow rate and also independent of the occurrence of air ingestion. The amount of entrained swirl is predetermined by the given submergence of the intake. Related to overall symmetrical conditions the entrained swirl decreases with increasing water depth (see Figure 1.10). If it is necessary to raise the water level in order to avoid air entrainment, then simultaneously swirl intrusion will be reduced. However, critical submergence conditions can not be directly related to critical swirl entrainment.

1.8. LAYOUT OF THE VOLUME

As mentioned previously, research and basic knowledge on the matter discussed in this monograph is still in progress. A complete understanding of all hydraulic phenomena involved in vortex formation and swirling flow problems at intakes is not yet available. But until now extensive basic and applied research has been carried out and numerous reports have been published. Important investigations were finished recently and some are still in execution and completion.

The subject under discussion is very complex and a state of the art report composed of summarizing contributions from a team of experienced researchers will therefore manifest the complexity of the whole matter by a survey of the numerous ideas, various opinions and different methods getting down to the solution of the apparent problems. This is thought to serve the main intention and aim of the volume in an appropriate way. Variety of the discussion will be preferred versus a more homogeneous representation of the technical material.

The subdivision of the volume into eight main chapters should cover the whole matter of concern giving special attention to considerations like: already available and generally accepted basic knowledge versus advances in science, approved tools and design methods versus new ideas suggested for application, or existing unsolved problems versus research intentions.

Chapter 2 (Fundamentals of vortex intake flow) summarizes available and approved basic information projecting it to advanced methods of interpretation and evaluation. In Chapter 3 (Basic results of theoretical and experimental work) the wide variety of the whole matter is narrowed down by concentration on general aspects and understanding of vortex phenomena at simple but basic geometrical and hydraulic conditions. Chapter 4 (Prediction of critical submergence) treats a very important feature of vortex formation comparing the state of present research results with the possibilities of an improved interpretation by concentration on some special aspects. With Chapter 5 (Modelling of vortices and swirling flows) an important excursion into the field of experimental techniques in the matter of concern is given showing the necessity of experimental work with respect to the necessary improvement of basic knowledge. Chapter 6 (Design recommendations for intake structures and for pump sumps) points out the general principles of vortex control and provides information on practical design procedures and application of research. In Chapter 7 (Vortex-flow intakes) desirable swirling flows in closed conduits such as dropshafts are treated and design recommendations are given. With Chapters 1 (Introduction) and 8 (Conclusions), the whole technical material is put in line with the envisaged purpose of the volume summarized by critical remarks on the state of the art and projected to further research needs.

Fundamentals of vortex intake flow 2

G.E.HECKER
Alden Research Laboratory, Holden, Massachusetts, USA

2.1. DESCRIPTION OF FLOW FIELD

2.1.1. *Analytical models of vortices*

A number of investigators have written the Navier-Stokes equations for a radially symmetrical, incompressible, and laminar flow field, and have derived solutions, using a variety of techniques and assumptions, to describe the velocity and surface elevation distributions. Due to their considerable complexity, these equations will not be presented here, but, more importantly for purposes of this volume, the basic nature and usefulness of some of these formulations will be briefly discussed.

The early simplifying assumptions of Einstein & Li (1955) may be cited as being typically used for analytical models, and these include radial symmetry, no vertical velocities nor vertical variations of horizontal velocities, hydrostatic pressure, and steady flow. The eddy viscosity concept is used, and turbulence is usually considered to increase the effective flow viscosity. Some of these assumptions, and the method of reducing the radial flow at the outlet, make predictions near a vortex core questionable. Using the same assumptions, Holtorff (1964) presented a numerical technique to better predict the surface depression and near core velocities.

The assumption of no vertical variations in horizontal velocities implies the absence of a bottom boundary layer. It is not very clear how the vorticity associated with such velocity gradients are stretched and amplified by vertical velocities near the core, but it is known that a (bottom) boundary layer does affect the final vortex phenomena. Intuitively, one can infer secondary currents produced by the vertical distribution of varying centrifugal forces. Anwar (1968) has conducted experiments showing that a vortex can be dissipated with a roughened bottom for flow conditions where a strong vortex would otherwise exist. Marris (1967) has shown analytically that a boundary layer would eventually cause reversal of the classical 'bath tub' vortex from an initial residual swirl opposite to the swirl direction produced by the earth's rotation. The presence of considerable vertical velocity variations near the core has been shown by Lewellen (1962), who also discussed the impact of ignoring the boundary layer.

Comprehensive analytical and numerical solutions are now available, such as those by Lewellen (1962) and Granger (1966), which apparently produce reasonably realistic descriptions of the vortex flow field in spite of some remaining simplifying assumptions. The difficulty in applying these analytical models to real cases, aside from the mathematical complexity, stems mainly from the requirement that reference values, or boundary conditions, of the flow field need to be specified. The validation of the available analytical models usually involves measurements in experimental facilities where the imposed circulation, or some analogous vortex flow property, is known and is used to solve for the remaining flow field, including water level variations. An additional difficulty in applying these analytical models is that the effects of local asymmetrical geometric features on the flow (including those of vortex suppressors) cannot, apparently, be analytically included. The difficulty with the available analytical techniques, therefore, is that they cannot predict if a vortex will (or will not) occur without considerable knowledge of the general flow field near the outlet. This knowledge must be gained using other independent techniques, such as intuition, physical (hydraulic) models, or other types of analytical flow predictors such as finite element (or difference) models (see Section 2.4).

2.1.2. *Rankine combined vortex*

As the equations of motion of vortex flow are rather complex and not amenable to a solution without knowledge of the surrounding flow field, a simplified analytical model seems justified. The typical schematic representation of vortex flow consists of assuming that a relatively small central portion of fluid is rotating as a highly viscous solid body ('forced' vortex), and that this is combined with a non-viscous 'free' vortex region extending radially outward therefrom. The two flow regions are usually matched at a radius proportional to the inlet radius, and for strong (air-core) vortices over a vertically downward inlet, the matching radius may be almost equal to the inlet radius.

In the central region of the vortex, the fluid is assumed to rotate so that the tangential velocity, v_t, varies linearly with the radius, r

$$v_t = \omega\, r \tag{2.1}$$

where

ω = angular velocity in radians per unit time.

In the free vortex region, the velocity varies inversely with the radius and directly with the circulation. By definition, the circulation of a vector flow field, $\vec{V}$, along any closed curve, C, is

$$\Gamma = \oint_c \vec{V}\, \vec{d\ell} \tag{2.2}$$

where

$\vec{d\ell}$ = elementary vector tangent to the curve C.

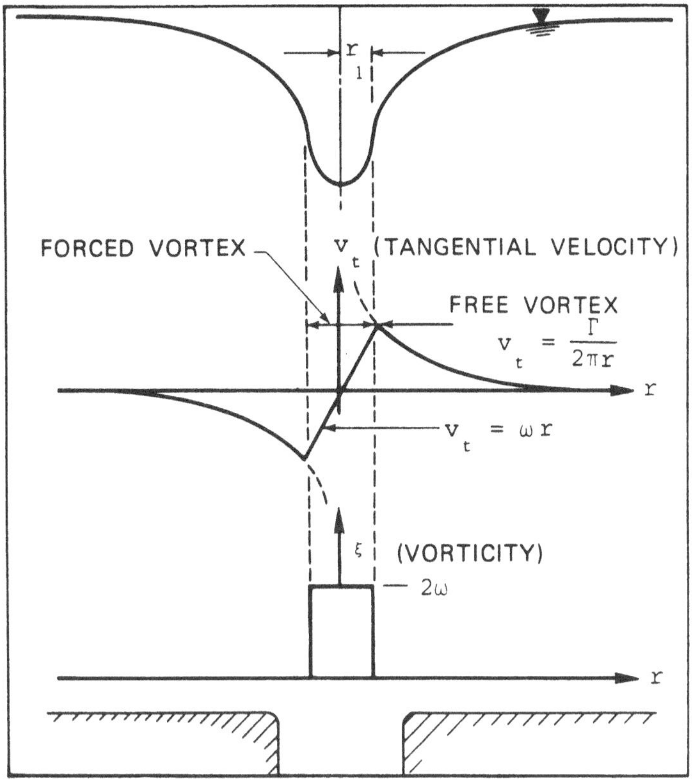

Figure 2.1. Combined vortex.

For a circular curve of radius r,

$$\Gamma = 2\pi r v_t \tag{2.3}$$

which is independent of the radius of C for a steady free vortex. Therefore,

$$v_t = \Gamma/(2\pi r) \tag{2.4}$$

Granger (1966) has shown that Equation (2·4) is an exact solution of the two-dimensional Navier-Stokes equations for axisymmetrical vortex flow. The variation of tangential velocity in the simplified matched model is shown in Figure 2.1, where r_1 is the radius at which the transition from forced to free vortex occurs. This radius is also the position at which the tangential velocity is maximum, although its actual value is affected by fluid viscosity, the latter also resulting in a smoothed transition of the tangential velocity variation with radius.

If the circulation around a closed curve is divided by the area within the curve, and the area approaches zero as a limit, it may be shown that

$$\lim_{\delta a \to o} \frac{\delta \Gamma}{\delta a} = \frac{\delta v}{\delta x} - \frac{\delta u}{\delta y} = \xi \tag{2.5}$$

where

ξ = vorticity component perpendicular to the plane of the elementary area, δa.

For zero vorticity, there is no rotation of fluid elements, and this is the case in the irrotational flow field of a free vortex.

Equation (2·5) indicates that the circulation around an area may be obtained by integrating the vorticity over the area. In the highly viscous forced vortex region, the vorticity is equal to 2ω, and the circulation at the matching radius, r_1, may be calculated based on the flux of vorticity (2ω) across the horizontal plane, πr_1^2, giving

$$\Gamma = 2\,\pi\,\omega\,r_1^2 \tag{2.6}$$

where no subscript is used on Γ since it is constant for $r \geq r_1$, i.e. in the free vortex region. Equations (2·3) and (2·6) are useful in calculating the circulation of a flow field for which the required variables may be directly measured or calculated analytically (see Section 2.4). Since there is only one constant circulation in the free vortex region, Equations (2·3) and (2·6) may be used to calculate the variation of the tangential velocity in the free vortex region as a function of flow conditions in the forced vortex, i.e.

$$v_t = \omega\,r_1^2/r \tag{2.7}$$

As mentioned in Section 2.1.1, there are a number of differences between this simplified description of the flow field and that of real intake vortices, including the neglect of vertical velocities and vertical variations in horizontal velocities. Also, in practical cases, local geometric features may generate vorticity which will influence the resulting circulation.

It may be seen that a major independent parameter in evaluating an intake's susceptibility to vortices is the magnitude of the approach flow circulation. For example, knowledge of the circulation allows determination of the maximum tangential velocity, v_{tm}, and Anwar (1966) has shown analytically that the depth of the surface depression at the vortex center, s, can be approximated by

$$s = 0.6\,v_{tm}^2/2g \tag{2.8}$$

The circulation in the approach flow may be influenced by features far removed from the intake, due to conservation of angular momentum, and it is frequently difficult to select boundaries (for analytical or physical models) where the flow characteristics are truly known. Proper simulation of boundary friction is also important, as the resulting velocity gradients will affect the cumulative circulation. Unsteady flow adds further complications in that the approach flow circulation is transported toward the intake, and may, therefore, vary with both space and time.

2.1.3. *Classification of vortex strength*

For a steady flow field and a given geometry, it may be assumed that the resulting flow over the intake will produce a particular maximum tangential velocity. The velocity head associated with the rotation will reduce the local hydrostatic pressure, resulting in a local drop in the water surface, as per Equation (2·8). This drop in the water surface may range from a so-called 'dimple' to, in the case of a strong circulation, an open air funnel or core the depth of the water column. The delineation of the various stages or strengths of vortices between these limits have been the subject of some debate as there is no applicable technical standard of differentiation.

A basic choice in classifying vortices is between visual techniques and the measurement of some quantity directly or indirectly dependent on the strength of vortices. For the latter case, changes in entrance loss coefficient (or discharge coefficient), the magnitude of inlet pipe flow swirl, or determination of the ingested air could be used. At least two problems arise, however, with this 'objective' approach. One is that even an air core vortex, if small compared to the inlet, may not produce effects that can reliably be measured, and the other is that the correlation between vortices and the selected dependent parameter may be weak (see Section 2.2.4) and variable with other parameters. Such measurements do, however, directly indicate the characteristics of the inflow and how these may affect a downstream use of the water. Measurement of variables directly related to vortices, such as the depth of the air core, are difficult as the vortex itself is usually not steady even for the presumed steady flow field. Such a method may also miss certain flow features of interest, such as the ingestion of floating trash into the intake. Similarly, a measurement of the circulation per se, although perhaps most useful, also does not by itself indicate flow problems.

Another way of evaluating the swirling flow field is to observe the type and strength of vortices being produced. Although this is clearly a subjective technique, a meaningful and consistent classification system helps considerably in evaluating a design and in comparing the performance of intakes. Such a visual classification method does not replace measuring, for example, inlet losses and inlet flow swirl, for such parameters are usually of interest in themselves. A visual classification system developed at the Alden Research Laboratory of WPI is based on the identification of distinct flow features indicative of a gradation of vortex strength. The basic idea is that stronger vortices are needed to ingest more buoyant material, the material varying from neutral density (dye), to floating trash, to air. Figure 2.2 shows this classification system, which indicates that a Type 3 vortex corresponds to a dye core, and is the demarcation between general swirl (Type 2) and concentrated vorticity from the free surface to the inlet. From Type 3, which involves a minimal surface depression (dimple), stronger vortices are identified until air bubbles are broken from the tip of the vortex air core (Type 5) and a full air core exists from the free surface to the inlet (Type 6). The vortex core (dye to air) may

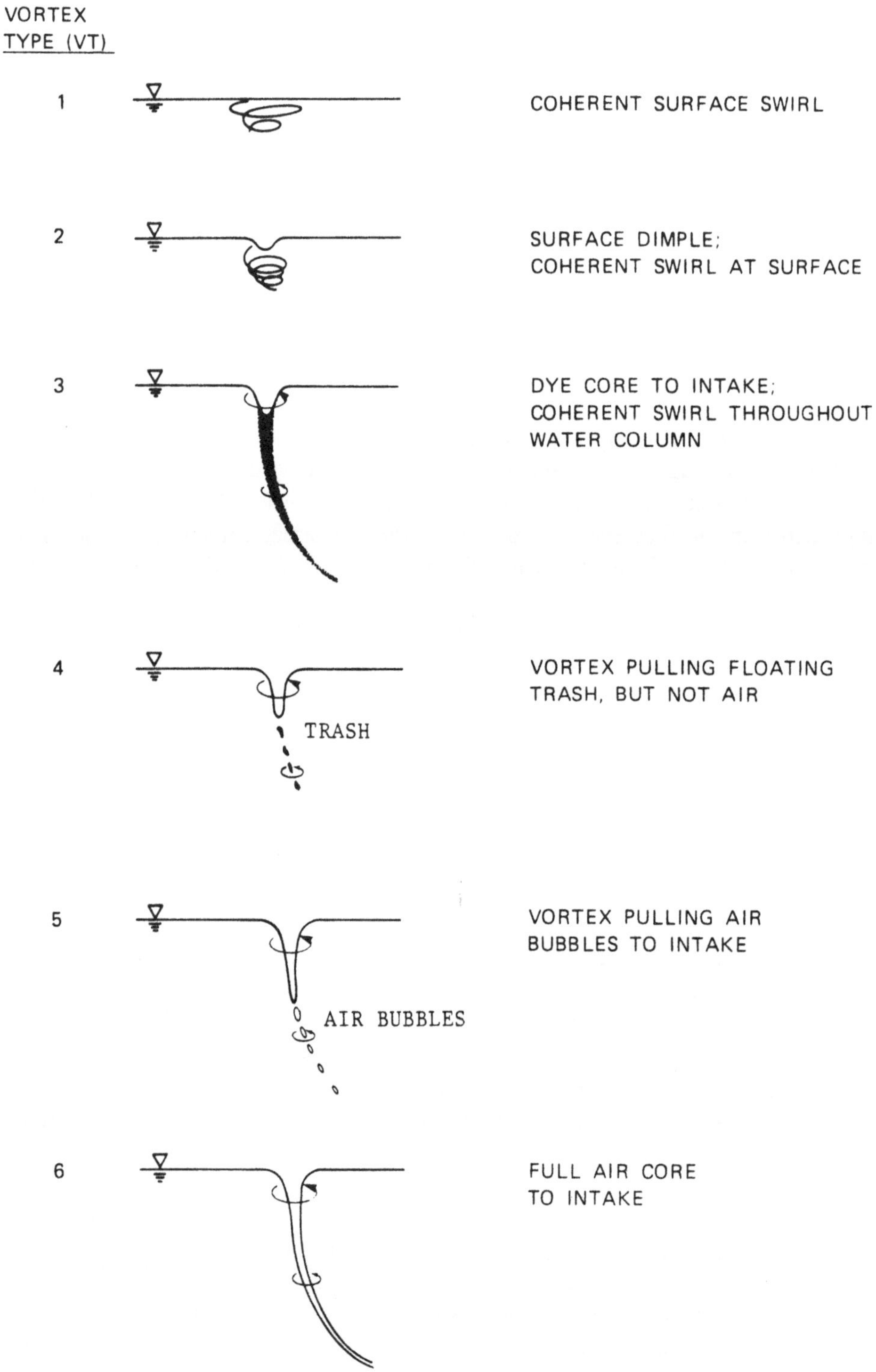

Figure 2.2. ARL vortex type classification.

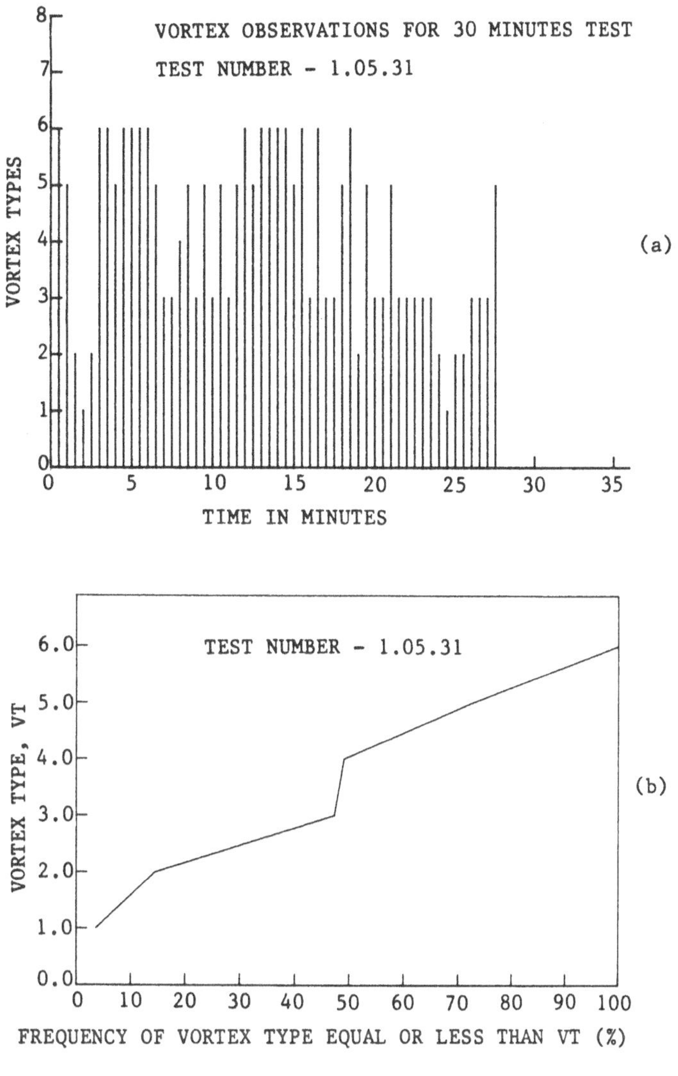

Figure 2.3. Time history and persistance of vortex types.

follow a direct or a tortuous path to the inlet, as influenced by the surrounding flow field.

2.1.4. *Time dependency of vortex strength*

Even for a steady discharge at fixed water level, a number of variable flow phenomena will produce unsteady vortices, the strength and position of which will vary with time. Small changes in approach flow distribution, varying turbulence intensity, small changes in flow separation from boundaries, and local variations in energy dissipation, all interact to produce unsteady vortices. Clearly, a strong air core vortex may appear stable, but most cases of interest to the practicing engineer will involve instances of borderline (weak) vortices. Once a vortex is reduced in strength or dissipated, such as by wall friction, it takes some time for the flow to 'reorganize' and produce enough circulation for the vortex to reform.

A careful dilineation of vortex flow, therefore, requires that the time dependency of vortex strengths (types) be recorded. One example of such a time series record is shown in Figure 2.3a. The advantage of such records is that they allow a careful definition of a characteristic vortex type for a given design, such as the true average vortex or, for example, the vortex type exceeded only 10% of the time. This avoids an arbitrary subjective association of a vortex type, such as the single worst vortex, with a given design. Figure 2.3b shows a persistence plot derived from the time series data of Figure 2.3a.

2.1.5. *Dependent variables – Dimensional analysis*

To allow the investigation of phenomena and the reporting of results in a manner which is applicable to any sized flow system without concern for the system of units, it is desirable to use dimensionless parameters. The most basic approach in determining which dimensionless parameters are of importance is to write the equations of motion for vortex flow in dimensionless form. In doing so, a number of dimensionless parameters will be seen to govern the flow. This approach has been used by a number of investigators, for example, by Lewellen (1962), and resulted in the identification of the following relevant dimensionless parameters:

$F_h = V/\sqrt{g\,h}$ = submergence Froude number
$N_\Gamma = \Gamma\, D/Q$ = circulation number
$R_R = Q/(\nu\, h)$ = radial Reynolds number
h/D = relative intake submergence

where

$V = Q/\pi\, D^2/4$
h = intake submergence
D = intake diameter
g = acceleration due to gravity
Q = intake discharge
ν = kinematic viscosity.

If boundary conditions for the equation of motion included surface tension forces, an additional dimensionless parameter would result:

$We = V\,\sqrt{\rho\, D/\sigma}$ = Weber number

where

ρ = fluid mass density
σ = surface tension force per unit length.

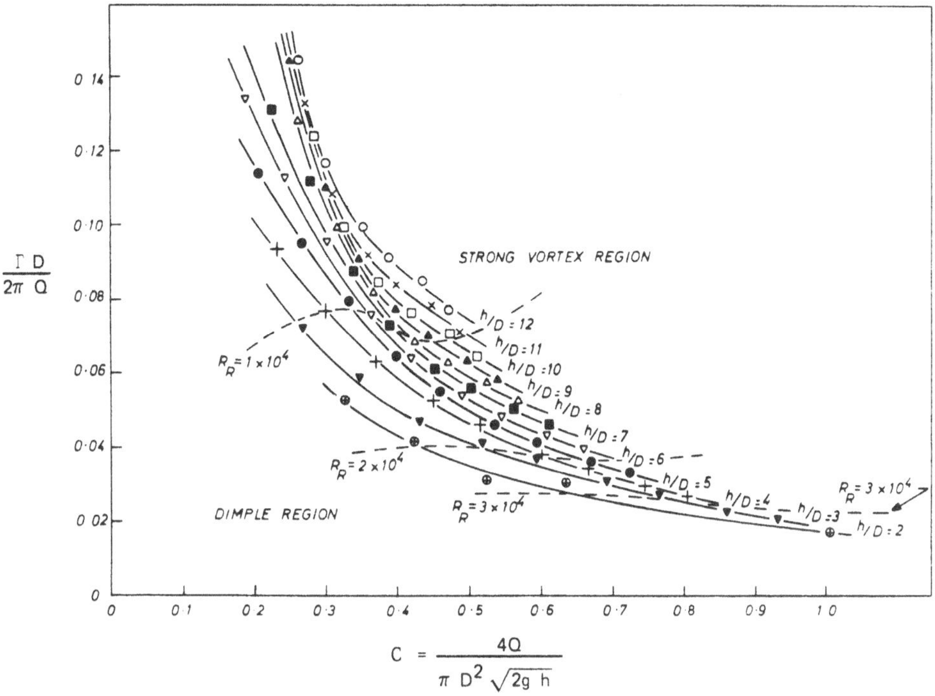

Figure 2.4. Vortex occurrence criterion for horizontal intake (Anwar et al., 1978).

Although a clear demonstration of the effects of surface tension is difficult to provide, tests by Anwar et al. (1978), Daggett & Keulegn (1974), and Jain et al. (1978) tend to show that surface tension has minimal effects on vortices for $We > 10^4$. The value of R_R above which vortex characteristics are essentially independent of viscous influences has been investigated by various researchers, and Anwar et al. (1978) has indicated that the critical value of R_R is about 3 10^4. For essentially all field applications, the critical values of We and R_R will be exceeded, and the dependency of vortex characteristics on We and R_R is mainly relevant in the applicability of hydraulic model and basic research results. Assuming We and R_R are sufficiently high so as not to have any significant impact on vortex characteristics, it may be stated that

$$VT = f\,(F_h, N_\Gamma, h/D) \tag{2.9}$$

where

VT = vortex type (see Figure 2.2)

f = function dependent on intake geometry.

One way to represent the function f is to give the values of F_h, N_Γ, and h/D, for which a particular type of vortex occurs, all for a given intake. Generally applicable data are available for simple samples, such as vertically downward, horizontal, or vertically upward withdrawals using round outlets with no appurtenant

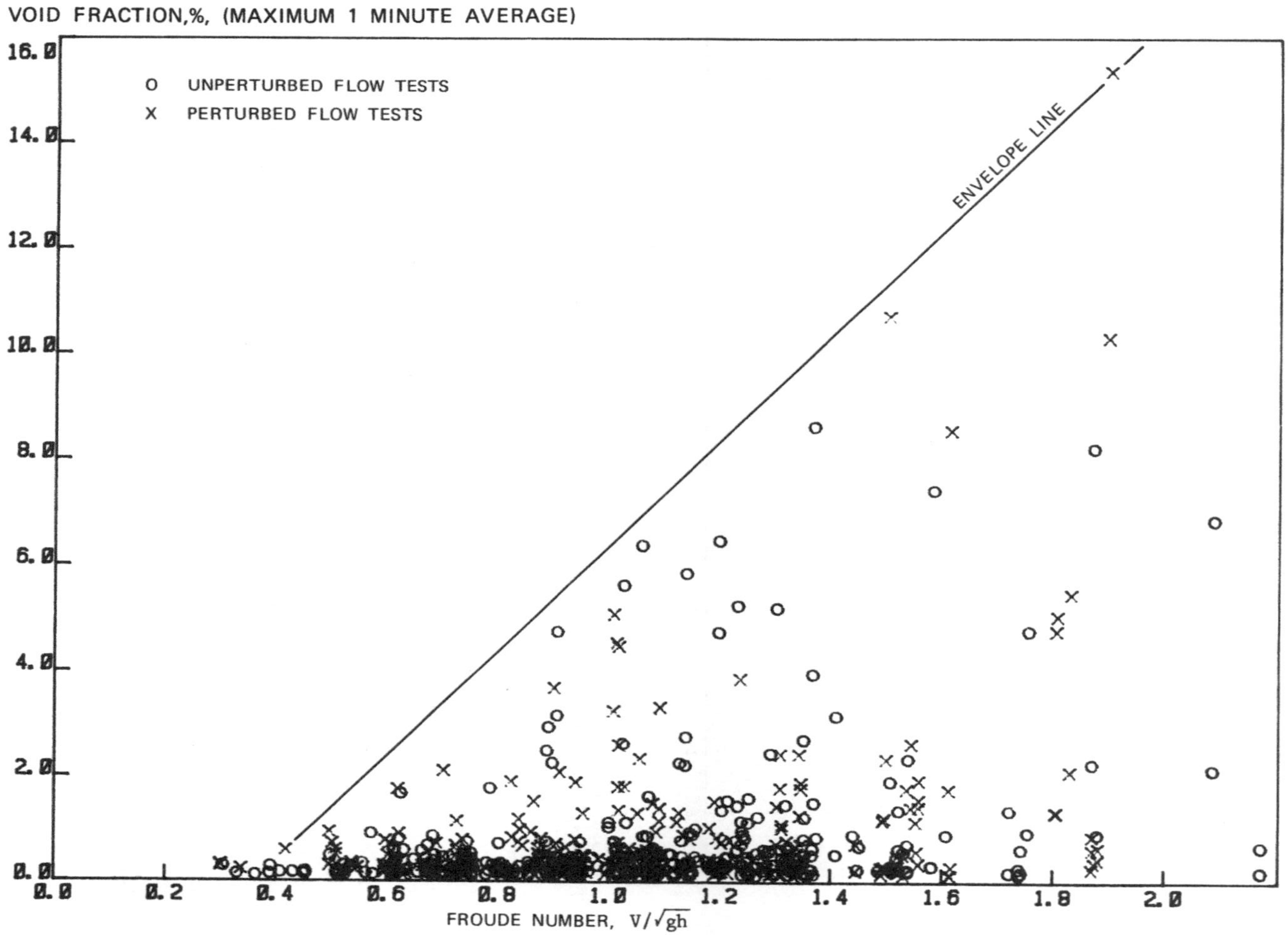

Figure 2.5. Envelope of maximum air ingestion versus Froude number (Padmanabhan, 1982).

structural elements. For example, Figure 2.4 shows such data by Anwar et al. (1978) for a vortex type for which the air core just reaches the horizontal intake, i.e. Type 6 on Figure 2.2. In turn, knowledge of the flow variables of Figure 2.4 for a given situation would allow determination as to whether an air core vortex would or would not occur. It may be appreciated that the most difficult parameter for a design engineer to determine is the circulation, which in many real cases is itself a function of the submergence due to changes in approach topography with water depth. In some experimental cases, it may be possible to generate the maximum circulation which might reasonably be expected in a given circumstance, and plot a vortex dependent parameter, such as air ingestion, versus the Froude number existing for each tested flow condition. Such a data set and envelope line are shown in Figure 2.5 for sumps with two horizontal (round) outlets situated below the approach floor level. Use of this figure does not require knowledge of the circulation. However, although yielding the maximum probable vortex characteristic, such data does not indicate to what extent the actual vortex characteristic could be below the maximum value for a given geometrically similar intake.

It might be noted that the intake diameter may be substituted for the intake submergence in the Froude number, although the free surface characteristics of this dimensionless number would then appear lost. Also, a number of investigators favor a Reynolds number based on the inlet pipe velocity and diameter, although this might better serve the study of such parameters as the intake loss coefficient. Other forms of the Reynolds number might prove useful for submerged vortices (see Section 2.1.6).

2.1.6. *Submerged vortices*

In many applications, submerged (or so-called 'wall' vortices) may be even more detrimental than free surface vortices, particularly when occurring in 'wet pit' pump intakes. Submerged vortices involve coherent swirl between a boundary (wall or floor) and an intake, or between two adjacent intakes. Unlike free surface vortex cores, the pressure in submerged vortex cores can be below atmospheric pressure in association with a higher value of circulation. If the pressure in such a vortex core is sufficiently low, dissolved gases will come out of solution, making the vortex core visible. The core can also be made visible by the injection of dye, an important procedure for model studies. Submerged vortices generate comparable or even more swirl of the inlet flow than produced by free surface vortices and, since both submerged and free vortices may occur simultaneously, the former may mask the swirl generated by only the free surface vortex. Figure 2.6 shows both types of vortices occurring simultaneously.

Proper reproduction and modelling of swirl intensity caused by submerged vortices requires minimal influence of viscosity. For depressed sumps, tests by Padmanabhan & Hecker (1984) have shown that a limiting approach flow Reynolds number,

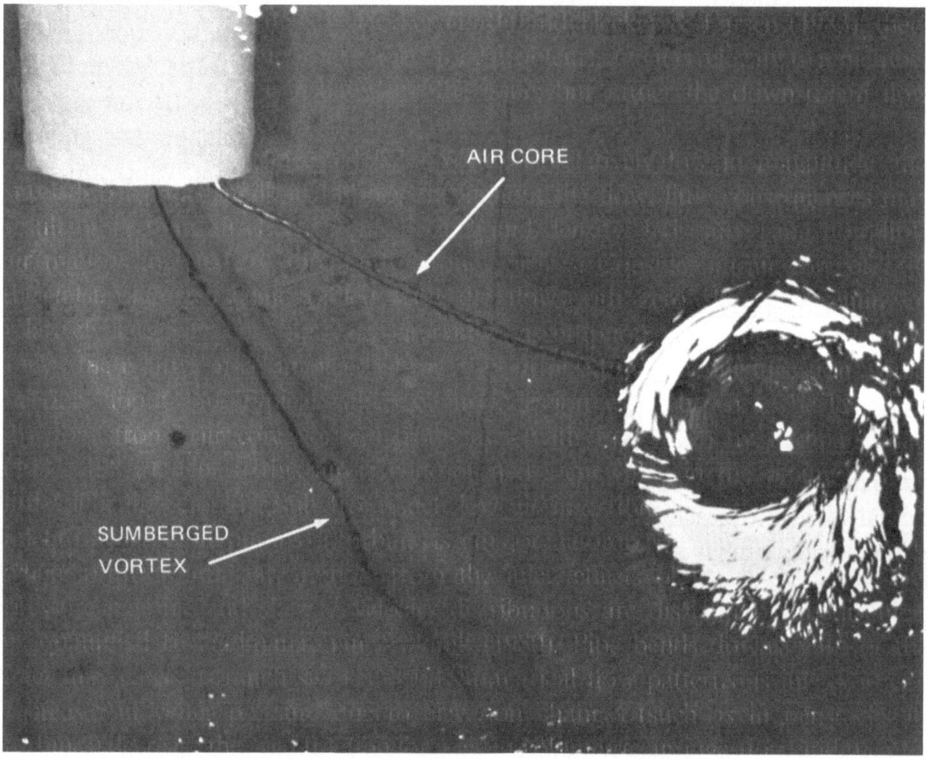

Figure 2.6. Plan view of horizontal intake with surface and submerged vortex.

$u_a s_a/\nu$, must be exceeded, where u_a and s_a are the approach velocity and depth, respectively. Figure 2.7 indicates this limiting Reynolds number has a value of about 3×10^4.

2.2. MEASURING TECHNIQUES

A short discussion of the various means by which vortex characteristics and their effects may be measured is of assistance in understanding the consequences of vortices.

2.2.1. *Vortex circulation*

As indicated by Equation (2.3), measurement of the tangential velocity at a known distance from the vortex center is sufficient to establish a fundamental vortex feature, the circulation. As only a weak disturbance may be sufficient to disrupt a small vortex, it is preferable that the velocity measurement involve minimal imposed drag. Various optical methods have been used in laboratory situations,

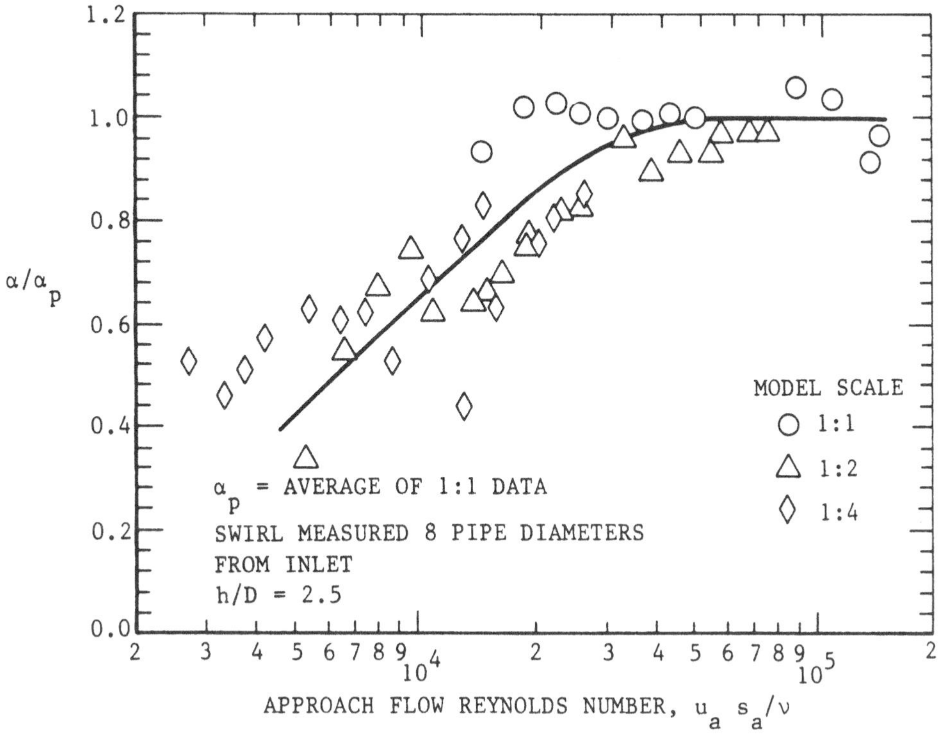

Figure 2.7. Effect of approach flow Reynolds number on inflow swirl angle (Padmanabhan & Hecker, 1984).

from timing the motion of neutrally buoyant particles in the water column to measuring the size of the shadow on the basin floor caused by the surface depression. This shadow size has been shown by Anwar et al. (1978) to be directly related to the circulation for given flow conditions.

For typical hydraulic models, it is easy to time the radial motion of small floating tracers, but the results may be questionable due to effects of surface tension and varying radii during motion of the tracer. A tetrapod device was adopted by the Alden Research Laboratory, consisting of four partially submerged floats rigidly connected to each other. This assembly may be easily centered over the vortex and the rotation rate determined.

It might also be noted that the angled vane devices used by researchers to generate circulation may not produce the desired or a known value of circulation. The resulting circulation, which depends on the features of the vanes, should be independently measured.

2.2.2. *Inflow swirl*

A direct and accurate determination of the swirling flow pattern in the outlet conduit may be obtained by measuring the tangential velocity component at

Figure 2.8. Swirl meter in model of pump intake.

numerous points at the conduit cross-section of interest. Such measurements will allow evaluation of complex swirling flow patterns, including possible double spirals, and appropriate reduction of the point values results in the average swirl and the flux of angular momentum. Simultaneous measurements of the axial velocity at each measuring station will also yield the local swirl angle, defined as

$$\alpha = \text{arc tan } v_t/v_a \qquad (2.10)$$

where

v_a = local axial velocity.

Since detailed velocity measurements are time consuming, and frequently only the cumulative cross-sectional average values of swirl are of interest, it is often sufficient to use a 'swirl meter'. This device, shown in Figure 2.8, is essentially a propeller with zero pitch, producing no rotation for flow with zero swirl. Based on measured tangential velocity profiles, the swirl meter diameter should be about 0.75 of the pipe diameter to occupy the 'solid body' rotational flow field. The average flow swirl angle may be determined from

$$\alpha = \text{arc tan } \pi \text{ D n/V} \qquad (2.11)$$

where

D = pipe (or swirl meter) diameter

n = revolutions of swirl meter per time.

The swirl meter will yield meaningful information if the distribution of tangential velocity is in general conformance with highly viscous, 'solid body' rotation. Variation from such a velocity distribution are somehow averaged out by the resulting drag on the swirl meter vanes, and large departures from solid body rotation will produce misleading results. For example, a vortex core passing asymmetrically through the swirl meter implies the presence of viscous and non-viscous flow regions, the latter having negligible vorticity, and this situation may produce little rotation of the swirl meter. It is preferable to locate a swirl meter sufficiently far downstream from an inlet so that zones of high shear and varying vorticity are dissipated to yield an overall flow swirl.

Both the local velocity data and the swirl meter rotations may be instantaneous or time average values. Although time average values are traditionally used, it should be noted that the time average value may be acceptably low, whereas large fluctuations in swirl may occur with time. In some instances, such as when machinery is involved, the instantaneous values are more relevant.

2.2.3. *Air ingestion*

The presence of air due to strong vortices may or may not have detrimental effects on the operation in question. For centrifugal pumps, Florjancic (1970) and Murakami et al. (1969) have shown that up to approximately 3% air by volume distributed in the suction line does not serious degrade pump performance.

In cases where the magnitude of air will be a determining factor, commercially available void fraction meters may be used to measure the percent of the cross-sectional area occupied by air. The maximum flux of air compared to water may be estimated by assuming the air moves at the same velocity as the water. A visual inspection of the distribution of the air in the cross-section assists in determining if only a portion of the velocity profile should be used for this calculation.

Studies by Padmanabhan (1982) have shown that typical values of the void fraction for seemingly strong air core vortices are between 1 and 2 per cent. Vortices leading to such air ingestion values had air core diameters at the inlet of about 0.1 of the inlet diameter, which may be seen to represent relatively large air cores. Many air cores are much smaller, producing void fractions much less than one per cent. The large air funnels produced at low submergences and high Froude numbers can, however, produce void fractions of more than 10% as shown by Figure 2.5.

2.2.4. *Inlet loss coefficient*

A measure of the severity of vortex flow at an inlet is the effect on entrance losses. Two mechanisms are presumably involved; one involving form losses produced by a change in flow patterns at the entrance, the other involving greater friction losses in the entrance region due to swirling pipe flow (see Section 2.3.3). The standard procedure for determining entrance losses is easily applied, and consists of measuring the pipe pressure gradeline over a sufficient length downstream from the intake to include the asymptotically achieved gradeline for uniform, fully developed pipe flow. Addition of the average velocity head and extrapolation to the intake face of the total energy line, at a slope equal to the fully developed gradeline, allows the head difference from the 'reservoir' water surface to be determined. Division by the velocity head yields the loss coefficient, which includes all entrance losses, commonly used for engineering design purposes.

Of interest here is how this loss coefficient is affected by vortices at the inlet. A clear dependency between these two factors may at first thought appear obvious, but extensive measurements by Padmanabhan (1982) have shown no statistically significant dependence between the vortex types of Figure 2.2 and the inlet loss coefficient. It must be noted that the vortex classification has as its maximum the initial occurrence of an air core to the inlet, and larger air cores occupying significant portions of the inlet area are not delineated. Relatively large air core vortices would have an effect on the inlet loss coefficient, as shown by Anwar (1966) and Daggett & Keulegan (1974). It appears, however, that vortices with 'incipient' air cores, or lesser types, do not sufficiently increase the inlet loss coefficient to warrant concern.

2.3. VORTEX CIRCULATION VS. INTAKE FLOW SWIRL

Although swirl is a prime factor in vortex flow, it does not constitute an engineering problem until contributing to increased entrance losses or swirling pipe flow downstream from the inlet. The potential safety hazard of strong circulation relative to small boats should, however, not be neglected.

2.3.1. *Detrimental effects*

It is clear that excessive circulation associated with vortices will reduce the coefficient of discharge of the inlet, or increase the inlet loss coefficient. However, except for basic studies in which the vortex circulation is measured, little information is available on the effect of circulation on entrance losses. The measurement of flow swirl in the inlet pipe has become more common, and some correlation between this swirl and entrance losses is available. Figure 2.9 shows some data by Padmanabhan (1982) for a horizontal outlet pipe from a depressed sump, the trend showing some dependence between inlet losses and the flow swirl angle, as defined in Equation (2·11) for use with a swirl meter. As both free surface vortices and submerged wall vortices contribute to the total measured swirl angle, the contribution of only free surface vortices is difficult to separate.

Swirl of the inlet flow may directly affect a downstream water use, and a common

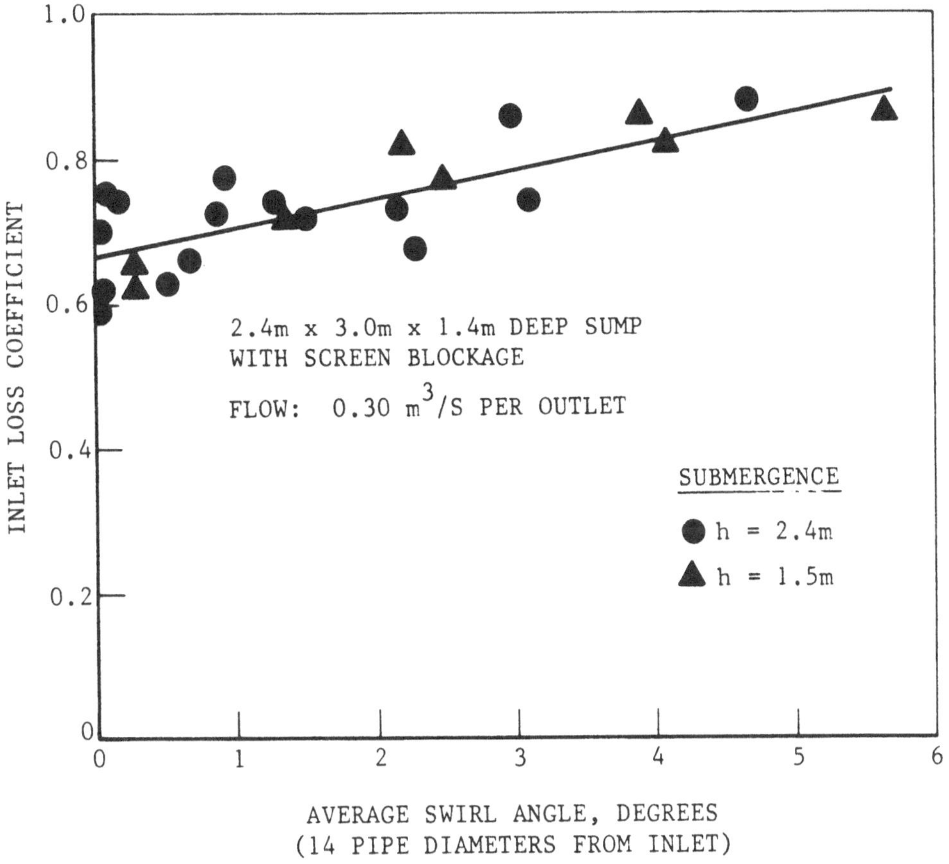

Figure 2.9. Effect of inflow swirl on intake loss coefficient (Padmanabhan, 1982).

concern is the effect on the performance of pumps and turbines. An analogous situation is the detrimental effects of conduit bends just upstream from pumps or turbines. Although little is known about the effect of pump inlet swirl on pump performance, it is generally accepted that the swirl angle just upstream of the pump should be less than about 5°. Larger swirl angles produce unfavorable angles of attack of the flow onto the impeller blades, producing excessive separation and loss of efficiency. In comparison to this criterion, measurements at the Alden Research Laboratory of the tangential velocity distribution just downstream from a standard, short-radius, pipe bend indicate an integrated swirl angle of about 8°.

An additional concern in some applications is the effect of swirl on the accuracy of flow meters. Padmanabhan & Janik (1980) show that flow with a swirl of 20° produces a 2% error in flow measurement using a standard ASME orifice meter with a diameter ratio of 0.6.

2.3.2. *Correlation of circulation and swirl*

As previously implied in Sections 2.2.2 and 2.3.1, the correlation between vortex type and inlet flow swirl is usually poor as the response of the swirl meter depends on the position of the vortex core relative to the meter axis, and submerged vortices may have a strong influence on swirl generation. This is particularly true for horizontal intakes for which the vortex (air) core usually enters near the top of the inlet. In some cases, such as with vortices situated directly over vertically downward inlets, the correlation between vortex types and inlet flow swirl has been good. A more basic correlation is between the circulation of a vortex and the inlet flow swirl, but this would require measurement of the tangential velocity distribution in the inlet. A more practical correlation of this type is that between the vortex circulation and the reading of a swirl meter located near the entrance of the inlet pipe. Using conservation of angular momentum between the vortex flow field to just inside the inlet pipe of radius R = D/2 and assuming solid body rotation and uniform axial velocity in the inlet, Whittles (1982) has shown that

$$\Gamma = \pi R^2 \omega_i \tag{2.12}$$

where

ω_i = angular velocity of the intake flow, as determined from the swirl meter rotation.

The actual distribution of tangential velocity in the inlet does not necessarily follow $v_t = \omega_i r$ and is probably related to the magnitude of swirl, among other factors. Also, the axial velocity distribution will not be uniform and was shown by Lee (1975) to vary with the swirl intensity. Therefore, it is more reasonable that

$$\Gamma = A\pi\, R^2 \omega_i \tag{2.12a}$$

with

$$A = A(\omega_i).$$

This expression indicates that the angular speed of rotation of a swirl meter, ω_i, can be used to approximate the circulation of the flow field upstream from the inlet if some data were available to estimate $A(\omega_i)$. Experiments by Whittles (1982) for a vertically downward round intake indicate that $1.0 \leq A \leq 1.4$, with no consistent trend between A and ω_i. Similar experiments with a horizontal intake by Padmanabhan (1982) for a nearly constant ω_i associated with steady air-core vortex indicate A is approximately 5.0. This high value might be influenced by the vortex filament passing asymmetrically through the swirl meter and thus causing a low rotation, although the swirl meter was located about eight pipe diameters downstream of the inlet and ω_i near the pipe inlet, calculated by the principles in Section 2.3.3, was used in the determination of A. The more likely reason for the high value of A is the presence of submerged vortices of opposite rotation to that of the free surface vortex. Although additional data is clearly needed, it does appear that knowledge of the vortex circulation gives some estimate of the result in pipe swirl, and vice versa.

A conservative approach would be to assume conservation of angular momentum, set A = 1.0, and calculate the inlet flow angular velocity, ω_i, using Equation (2.12). Again we see the importance of knowing the vortex circulation, Γ.

2.3.3. *Decay of inlet swirl*

Swirl produced by vortex flow will decay with distance along the outlet pipe due to shear at the pipe wall. As the tangential velocity increases with distance from

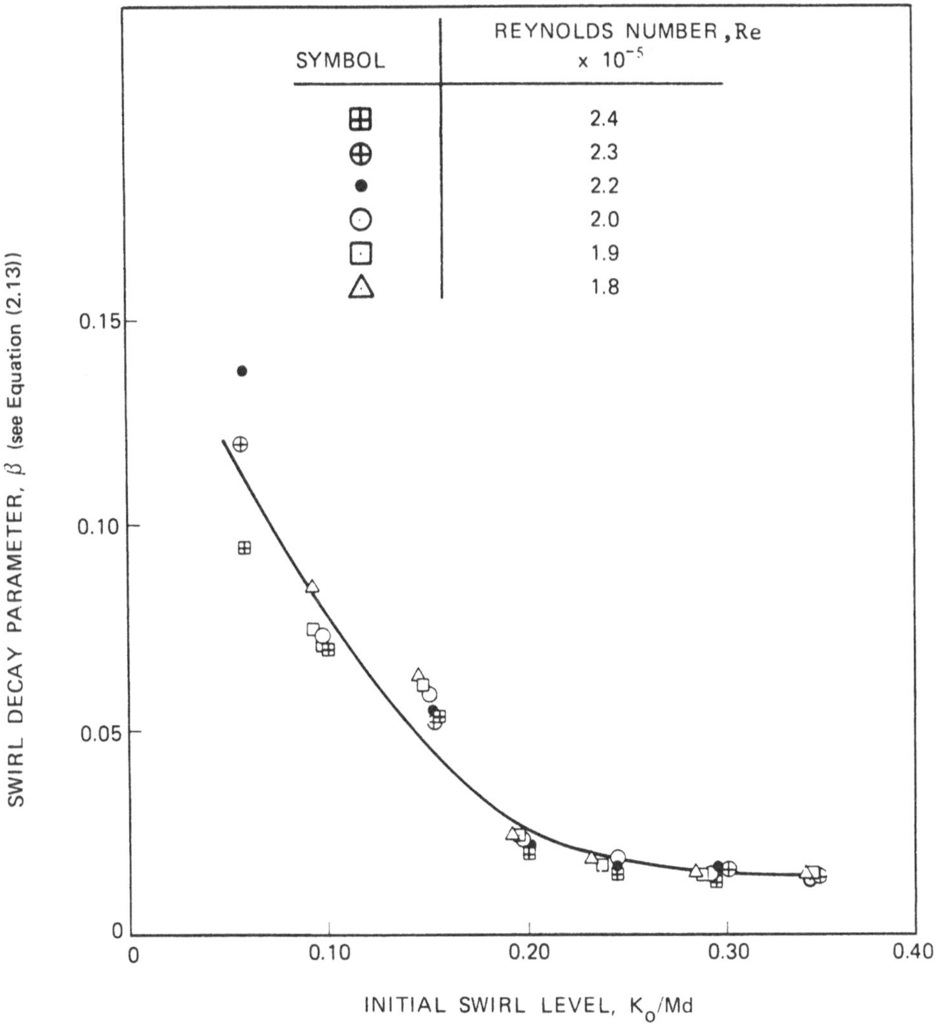

Figure 2.10. Dependence of the swirl decay parameter on initial swirl level (Padmanabhan & Janik, 1980).

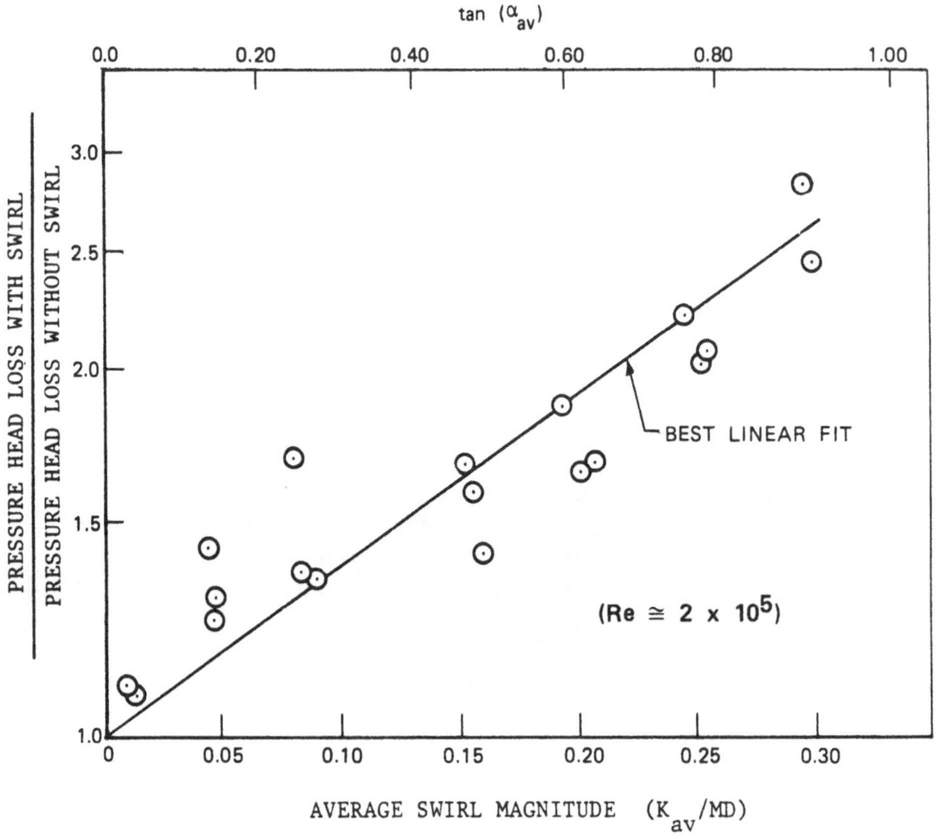

Figure 2.11. Effect of swirl on pressure head losses (Padmanabhan & Janik, 1980).

the pipe center, approximately in accordance with $v_t = \omega_i r$, wall friction acts on the near maximum tangential velocities, producing maximum dissipating forces. Lee (1975) showed that the axial velocity distribution is also altered by swirl to produce a velocity defect near the pipe center and maximum values near the pipe wall. Therefore, pipe friction is similarly effective in counteracting the distorted axial velocity profile. A knowledge of the decay of swirl with distance is important in determining possible effects of vortex produced swirl on downstream uses of the water, such as the effects on pump and turbine performance.

The swirl decay between two locations in a pipe an axial distance Δx apart has been shown by Baker & Sayre (1974) to follow an exponential function, which may be written as

$$\frac{K}{K_o} = e^{-\beta \Delta x / D} \tag{2.13}$$

where

K_o, K = upstream and downstream angular moment fluxes, respectively

β = decay parameter which depends on the pipe Reynolds number as well as on the initial upstream swirl.

The angular momentum flux, K, may be approximated by

$$K = \rho\pi R^2 V v_t R = \rho\pi R^4 V \omega_i \tag{2.14}$$

A non-dimensional swirl intensity was defined by Baker & Sayre (1974) as the ratio of the angular momentum flux to the axial momentum, $M = \pi\rho V^2 R^2$, i.e. by K/MD. Tests by Padmanabhan & Janik (1980) have shown that the swirl decay parameter, β, is a function of the initial (upstream) swirl level, K_o/MD, as shown in Figure 2.10.

The previously mentioned distorted velocity profiles result in more head loss than is the case without swirling flow, and this is a point to remember when making important head loss calculation. Figure 2.11, by Padmanabhan & Janik (1980) shows the pressure drop per unit of pipe length would be about twice as large for a swirl angle of 30° (K/MD $\cong$ 0.2) as compared to the pressure drop without swirl.

2.4. ANALYTICAL MODELING OF CIRCULATION

As has previously been mentioned, determination of the circulation around an outlet is a primary variable in describing vortex flow, in predicting the occurrence of vortices, and in estimating the amount of swirl present in the outflow conduit. Knowledge of the circulation, in conjunction with available analytical models described in Section 2.1.1 would allow analytical description of the complete vortex flow field. It is therefore worth considering how the circulation may be estimated and quantified, even though this represents a considerable challenge.

A direct and relatively accurate prediction of circulation is possible using a physical model, as long as proper attention is given to scaling sufficient approach geometry, the boundary roughness, and certain minimum Reynolds (and Weber) numbers are achieved. It is clear, however, that once a hydraulic model is available, it is more efficient to use the model to predict the possible occurrence of vortices and to determine the pipe flow swirl.

In some cases of relatively simple geometry, it may be possible to make reasonable estimates of the circulation. If the deviation of the streamlines from radial inflow can be intuitively approximated, the associated geometry may be sufficient to calculate an approximate value of circulation according to Equation (2.3). An example of this procedure has been given by Anwar (1968). In most cases, however, intuitive prediction of streamlines becomes rather unreliable. A promising analytic technique to determine circulation is the use of two dimensional numerical models which solve the depth-integrated conservation of mass and momentum equations through Finite Difference or Finite Elements techniques. The following describes the application of a particular two dimensional finite elements model to solve for the circulation around an outlet in a simulated reservoir.

2.4.1. *Finite Element modeling of circulation*

The Finite Elements Method (FEM) is a technique in which the flow field of interest is divided into a grid of elements, the spacial distribution and proportions of which are varied to suit the boundary shape and desired resolution of the flow in space and time. The depth average conservation of mass and momentum equations are formulated using mathematical techniques which transforms the partial differential equations into ordinary differential equations in time. Initial values must be prescribed for velocities and water depths at all nodes, and zero velocities and a horizontal water surface are convenient initial conditions. At solid boundaries, zero normal (perpendicular) velocities must exist, although the tangential velocity component may be either zero or unprescribed. If the boundary moves with time, such as for pumped storage reservoirs, node positions may be altered, and elements added or eliminated, with time.

Results of the FEM are velocities and water surface elevations at all nodes for each solution time step, and interpolation may be used to calculate velocities and water levels at any other location. Once the velocities are known at any given time, the circulation around an outlet may be calculated along a closed path consisting of element sides. Using Figure 2.12, the circulation may be calculated, based on Equation (2.2), by summing the tangential velocity and path components

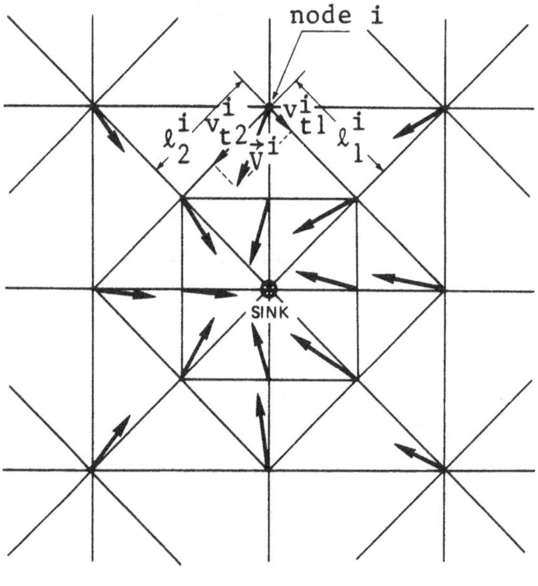

Figure 2.12. Calculation of circulation with finite elements grid.

(maintaining a sign convention) over the nodes involved, i.e.

$$\Gamma = \sum_{i} \frac{v_{t1}^{i}\ell_{1}^{i} + v_{t2}^{i}\ell_{2}^{i}}{2} \tag{2.15}$$

For the study by Brocard et al. (1982), used as example herein, the Wang & Connor (1975) model was used and the following comments apply mainly to this particular application.

In formulating the conservation equations for turbulent flow, the turbulent eddy viscosity approach is used to replace the so-called Reynolds stresses. The choice of eddy viscosity, ϵ, used in the FEM is an important factor in that its value should change depending on flow conditions (e.g. velocity gradients), but too small a value may lead to insufficient damping of numerical instabilities. The optimum value of the eddy viscosity near the sink(s) was selected by numerically solving for the circulation as a function of radius and time in a simulated circular tank with a known inflow circulation. As analytical models are available for such a case (see Section 2.1.1), a comparison between the analytical and numerical methods leads to the optimum eddy viscosity to be used in the FEM near the sink. The most reasonable reproduction of circulation in the simulated circular tank was achieved by Brocard et al. (1982) when

$$\frac{2\pi\epsilon h}{Q} = 0.5 \tag{2.16}$$

It might be noted that calculations far from the sink (see Section 2.4.2) require a higher value of the eddy viscosity to achieve numerical stability.

It was also necessary to distribute the sink flow over several nodes and adjust the outflow ratio among the nodes to avoid numerical instabilities. As long as the sinks are distributed uniformly along a given radius, their distribution would not affect the approach flow pattern.

2.4.2. *Application*

The ability of the FEM to determine proper values of circulation was tested by Brocard et al. (1982) by simulating the withdrawal from an idealized reservoir and comparing the calculated values to values of circulation measured in a hydraulic model of the same reservoir. For this case, most of the resulting circulation is generated far from the intake by non-uniform friction and the non-symmetrical bathymetry of the reservoir. Since some time of travel is involved from a location in the reservoir to the sink (outlet), the circulation grows with time and will vary depending on the distance from the outlet.

Figure 2.13 shows the simulated 1600 m 2400 m by 20 m deep (maximum) reservoir, the 500 m^3/s outlet location, and the finite elements grid used to calculate velocities and circulation. The 240 elements varied from 280 m to 25 m on a

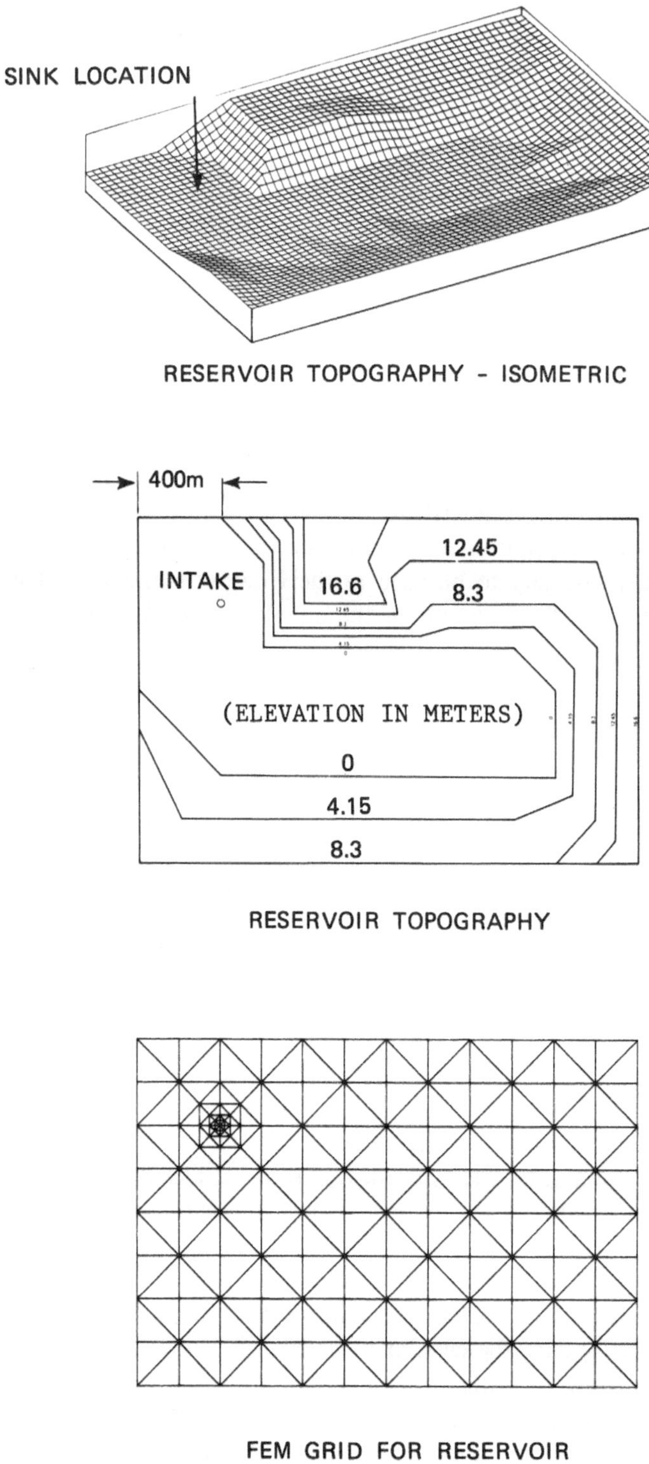

Figure 2.13. Simulated idealized reservoir (Brocard et al., 1982).

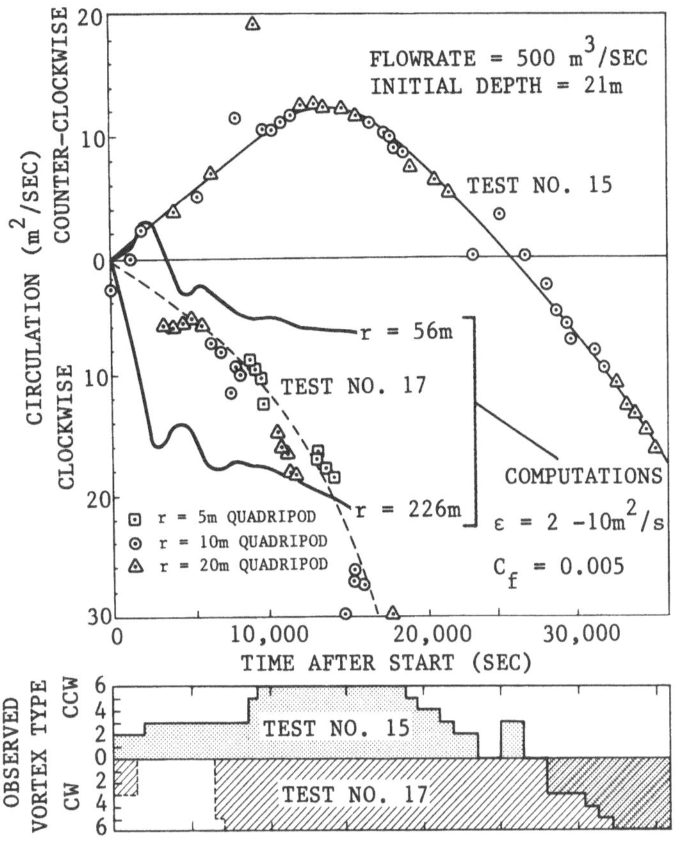

Figure 2.14. Comparison of calculated and measured intake approach circulation (Brocard et al., 1982).

side, requiring a solution time step of 1.25 seconds. The eddy viscosity near the outlet was 2 m^2/s, based on the previously mentioned criterion. Farther from the outlet, a value of $\epsilon = 10$ m^2/s was used, to avoid numerical instabilities.

The comparison between the measured and calculated circulations as a function of time, is shown in Figure 2.14. It is seen that the numerical calculations do not compare well with the actual circulation measured in the hydraulic model for the time period shown. This may be due to several reasons, including instabilities in the flow pattern which on occasion, gave considerably different results in the hydraulic model (as shown in Figure 2.14), and the inability of the numerical model to operate at sufficiently low values of the eddy viscosity. As a result of using high values for ϵ farther from the outlet, the Reynolds number of the calculated flow is low, influencing such phenomena as flow separation. The numerically predicted circulation was also found to be sensitive to the grid selection and the bottom friction coefficient.

It should be recognized that this is a severe test for a two-dimensional numerical model compared to other published applications, as the tangential components leading to circulation are relatively small compared to the total velocity vector.

Indeed, the calculated overall flow pattern to the sink appeared reasonable. Also, comparisons of circulation versus time is a more critical test than comparison of steady state values.

The above comparison illustrates, however, the difficulty in accurately calculating the value of circulation versus time at an inlet. Although improvements in numerical techniques will be made in the future, it appears that such techniques are presently not as reliable as hydraulic models for determining circulation and whether or not a vortex will occur.

Basic results of theoretical and experimental work 3

E.CHANG
Training Centre for Small Hydro Power, Hangzhou, China

M.J.PROSSER
BHRA Fluid Engineering, Cranfield, UK

3.1. INTRODUCTION

The problem of swirling flow at intakes has been the subject of numerous papers for many years. Swirling flow occurs in a wide variety of situations, e.g. as eddies shed from wing tips; in various atmospheric phenomena such as tornadoes and waterspouts; at the intakes to gas turbines operated near a ground surface; in pump sumps and intakes for hydro-electric schemes; in the common bath-tub and tea cup. Consequently, a wide variety of geometrical configurations and flow conditions have been used for both experimental and theoretical investigations.

It is this ease of producing swirling flow in widely varying situations which has led to conflicting opinions on the subject and misinterpretation of results. In order to narrow down the field of discussion, this section concentrates on basic aspects of vortex formation in a cylindrical tank with a drain outlet in the centre of the tank bottom (Figure 3.1). An initial swirl or circulation may or may not be superimposed. This arrangement produces relatively stable free-surface vortices and, together with the inherent axial symmetry, greatly simplifies both the experimental and theoretical treatment. Thus, the approach adopted here is to provide the reader with a basic understanding of vortex phenomena.

More information on practical applications to other geometries such as rectangular pump sumps can be found in Chapters 6.1 and 6.2 and in Chang (1977). Related aspects such as the origin of vorticity and critical submergence are treated in Chapters 2 and 4. Recent work on the mechanism of inlet vortex formation is reported in De-Siervi et al. (1982). Mathematical analyses are not specifically reviewed here, but brief details may be found in Chang (1976). They have in general been hampered by lack of reliable experimental data on velocities in the vortex core, and by the complexity of the basic equations when realistic boundary conditions are imposed.

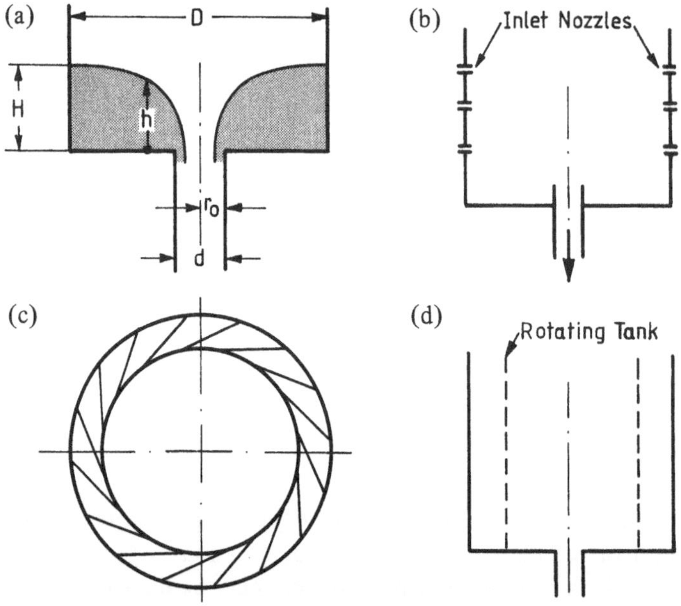

Figure 3.1. Experimental configurations.

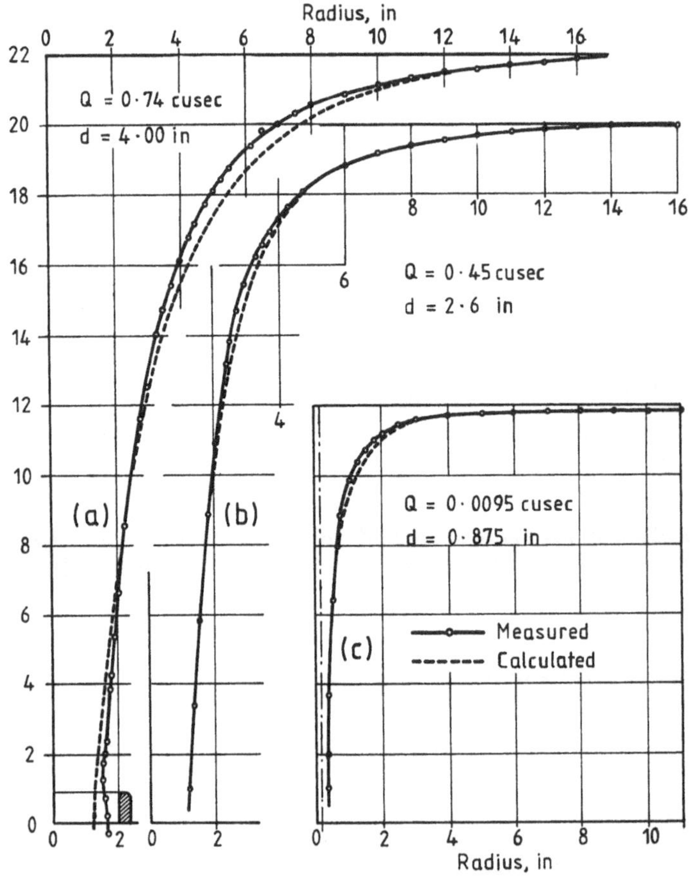

Figure 3.2. Measured and calculated profiles for: (a) large outlet pipe with high swirl and outlet projecting into tank, (b) large outlet pipe with high swirl and outlet level with bottom of tank, and (c) small outlet pipe with low swirl and outlet projecting into tank.

3.2. GEOMETRIC PROPERTIES OF THE VORTEX

3.2.1. *Surface profile near the vortex core*

Figure 3.2 shows the profiles obtained by Anwar, 1965 for strong vortices. The theoretical curves are based on inviscid theory, i.e. they are hyperbolic in form, but it can be seen that as the vortex core is approached, departures from the ideal become more significant. The measured profile is generally higher than calculated, and this is confirmed also by Stevens & Kolf (1959).

Deviations are to be expected due to two opposing factors: (a) the depth of liquid h, is not a constant near the axis of rotation and there is thus a reduction of radial flow area for the discharge. To satisfy continuity, this leads to higher exit velocities and thus a lowering of the surface; (b) as the velocity in the core increases, viscous shear forces also inrease, thereby reducing outlet velocities and increasing the water depth.

For weak vortices, Anwar (1966) has also reported that the shape is mainly cylindrical over the vortex length.

3.2.2. *Depth of surface dimple or vortex core*

Polikovski & Perelman (1956) and Anwar (1966) both give results in the form:

$$\text{depth } h \propto \frac{v_t^2}{2g}$$

where

v_t = tangential velocity component.

Figure 3.3 shows some typical results. Woodford (1964) found that the length of the air core increased with increasing imposed circulation Γ_∞ (constant circulation at large distance where it no longer varies with r). Also, for large water depths (H/d ≅ 10), the length was independent of suction velocity. However, at high values of imposed circulation, the length became almost independent of Γ_∞ but increased with increasing suction velocity (Figure 3.4). Woodford derived an

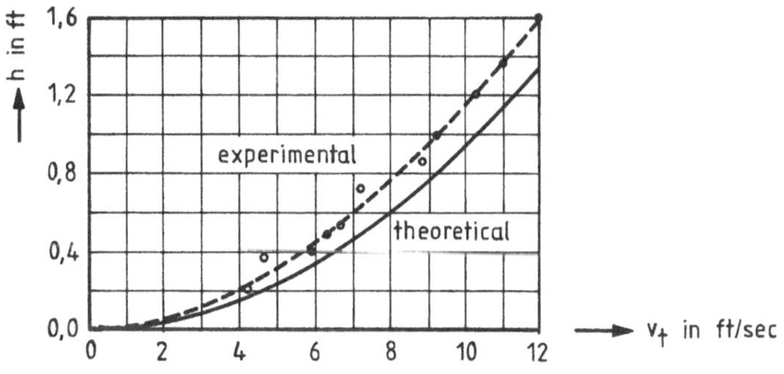

Figure 3.3. Variation of depth of dimple with change of tangential velocity v_t at outlet radius.

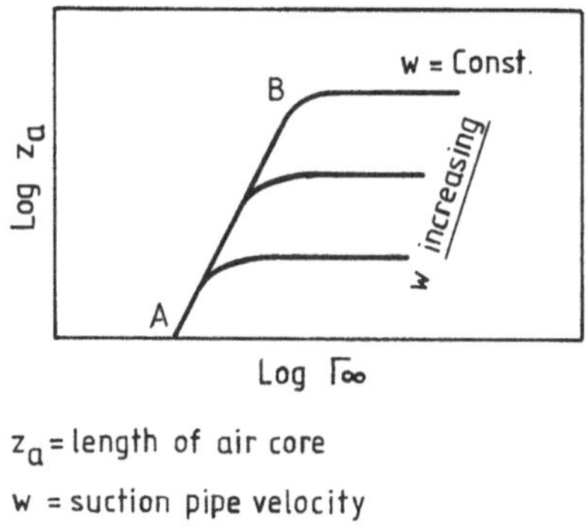

Figure 3.4. Qualitative graph of log Z_a vs. log Γ_∞.

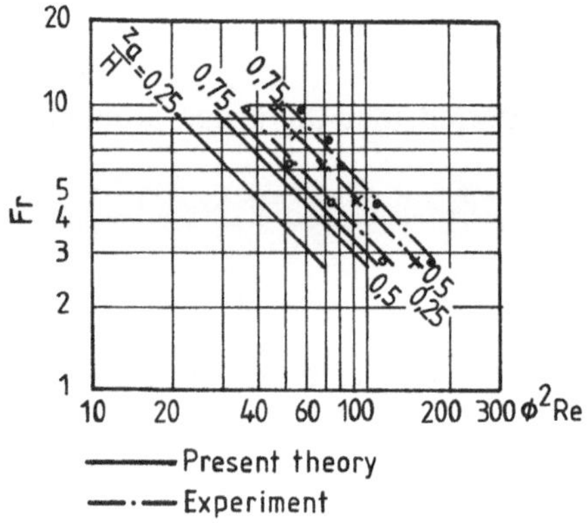

Figure 3.5. Fr vs. ϕ^2Re for air core lengths z_a/H = 0.25, 0.5, 0.75.

expression for the length of the air core as a function of

$$Fr\phi^2Re$$

where

Fr, Re are Froude and Reynolds numbers based on outlet diameter and velocity, and

$$\phi = \frac{\Gamma_\infty}{wd}$$

but obtained only qualitative agreement with experiment (Figure 3.5). Woodford used the cylindrical polar coordinates r, ϕ, z and defined u as the radial and w as the axial velocity component.

3.3. DISTRIBUTION OF CIRCULATION AND VELOCITY

3.3.1. *Distribution of circulation* $\Gamma = 2\pi v_t r$

Theoretically, the circulation Γ should be constant for any distance r from the

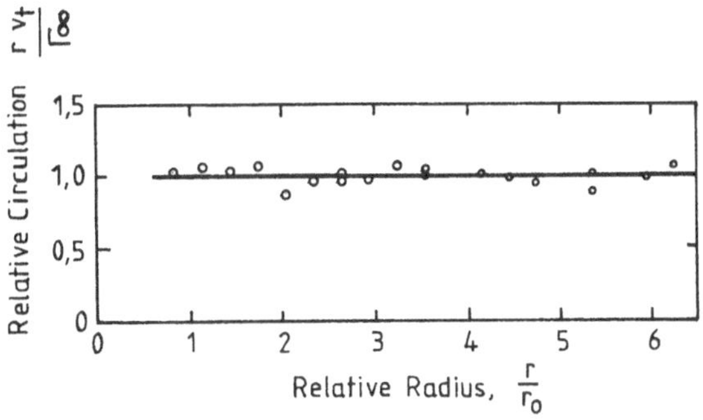

Figure 3.6. Variation of rv_t with distance from outlet axis.

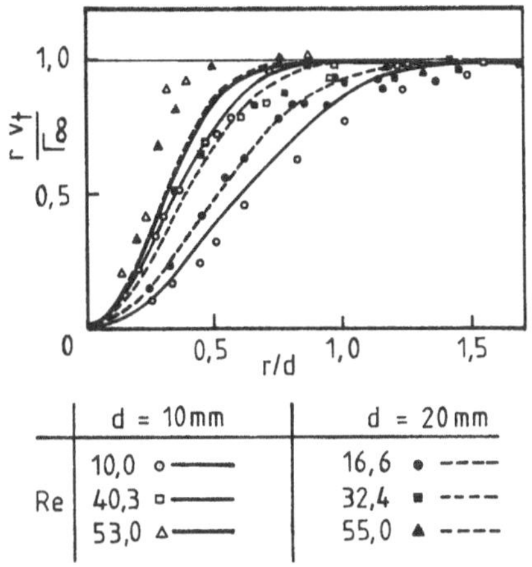

Figure 3.7. Decline of tangential velocity component and Γ near vortex axis.

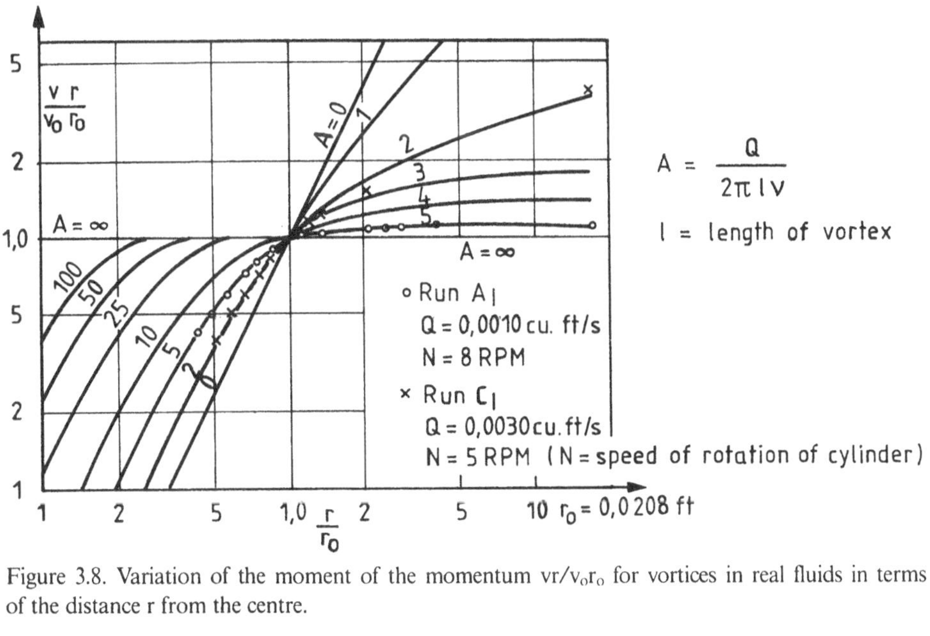

Figure 3.8. Variation of the moment of the momentum vr/v_0r_0 for vortices in real fluids in terms of the distance r from the centre.

outlet for a free vortex in an ideal fluid. This seems to hold true for the region outside the air core of the vortex (Figure 3.6) but for the region near the core, Toyokura & Akaide (1970) found that the circulation decreased to zero (Figure 3.7). The curves drawn in Figure 3.7 were obtained from theoretical considerations.

Einstein & Li (1955) investigated the effect of viscosity theoretically and obtained agreement with experiment over a limited range of variables with the assumption that turbulence in the core contributed an additional eddy viscosity to the total effective viscosity. It can be seen from Figure 3.8 that for large values of A, i.e. viscosity tending to zero, the circulation is almost constant for large r, but decreases rapidly as r decreases, showing the effect of turbulence in the core. The lines A = O and A = ∞ correspond to solid body rotation and an ideal free vortex respectively. A is effectively a kind of radial Reynolds number.

Anwar (1968a) found by dimensional analysis that

$$\frac{\Gamma}{\Gamma_\infty} = f\left(\left(\frac{r}{r_o}\right)^2, \frac{Q}{\nu H}\right)$$

with no dependence of circulation on water depth. From theoretical considerations, this simplified to

$$\frac{\Gamma}{\Gamma_\infty} = \left(\frac{r}{r_o}\right)^{2n} \quad n < 1 \text{ and } n = f\left(\frac{Q}{\nu H}\right)$$

Figure 3.9 shows the variation of n with radial Reynolds number $\frac{Q}{\nu H}$. The experimental points were obtained using glycerol solutions of kinematic viscosity in the range 11 to 40 times that of water. As viscosity increases, so n → 1 which corresponds to solid body rotation. Similarly, for high radial Reynolds numbers n → 0, corresponding to a free vortex.

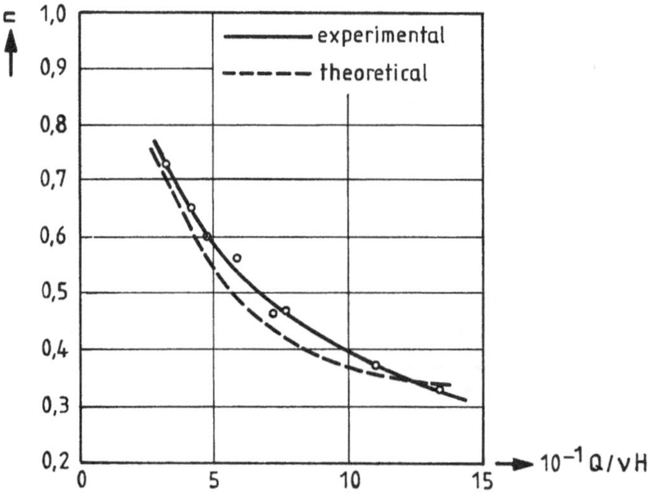

Figure 3.9. Variation of non-dimensional parameter $Q/\nu H$ with exponent n.

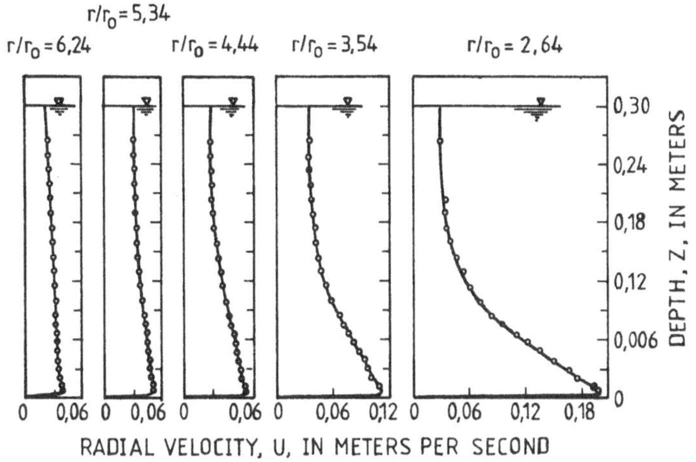

Figure 3.10. Radial velocities versus depth.

3.3.2. *Distribution of radial velocity u*

This velocity component is very small near the core and thus difficult to measure accurately. Daggett & Keulegan (1974) found that u varied not only with r but also with depth z, as shown in Figure 3.10. This is contrary to assumptions made in many theoretical analyses that

$$\frac{du}{dz} = 0$$

and, in fact, the radial velocity can become very high near the vortex core. Toyokura & Akaide (1970) also found that values of u at the free surface increased as the vortex core was approached.

3.3.3. *Distribution of tangential velocity v_t*

Anwar (1965) has found the distribution to follow that of an ideal free vortex (hyperbolic) except near the vortex core (Figure 3.11). Rohan (1965) has derived an expression for v_t near the air core in a simple power law from:

$$v_t r^x = \text{const}$$

Rohan found that for water, x had a value between 0.6 and 0.8. Anwar also found that v_t does not depend on z, thus substantiating assumptions made in theoretical considerations. Measurements by Daggett (Figure 3.12) also show that v_t is independent of z except near the tank bottom where boundary layer effects retard the flow.

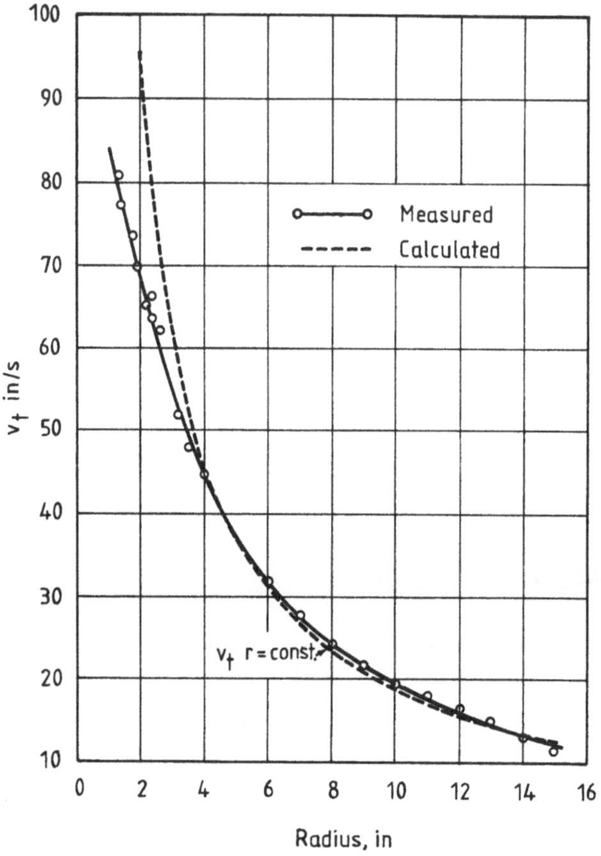

Figure 3.11. Comparison between calculated and measured tangential velocities far from core.

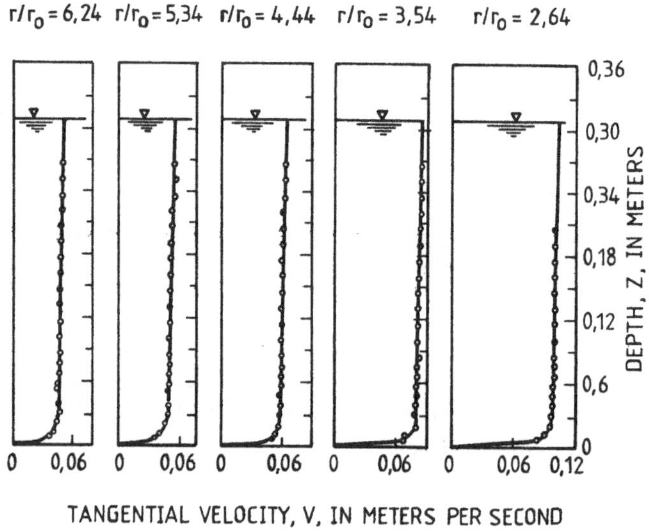

Figure 3.12. Tangential velocities versus depth.

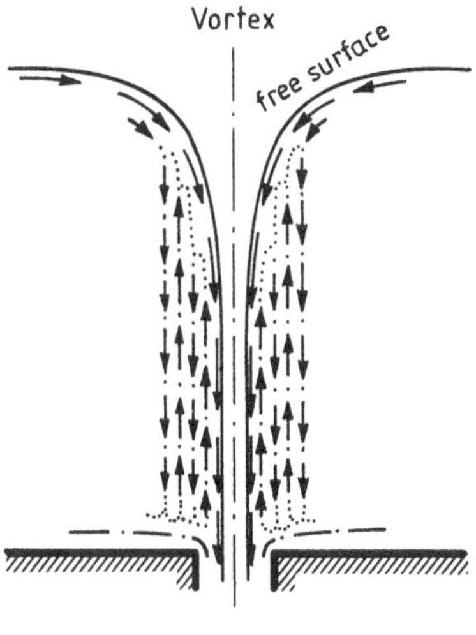

Figure 3.13. Sketch of observed radial and vertical velocity pattern near core of vortex.

3.3.4. *Distribution of axial velocity w*

This velocity component is small in the main body of liquid, though Quick (1962) has observed values of up to 10% of v_t. Upward as well as downward flows have been observed by several authors, as well as predicted theoretically (Figure 3.13).

Kamel (1964) has shown that these upward axial components probably occur where the flow through the boundary layer formed at the tank bottom exceeds the discharge capacity of the outlet, and so is forced upwards. Anwar (1966) has observed w to be practically zero on the axis of symmetry, and to have a maximum value at a radius of 0.75 r_o. Woodford (1964) also observed this maximum, with a central region in the vortex core where w was lower than at larger radii. This central region occupied a decreasing proportion of the region of downward axial velocity as the product ϕ^2Re was reduced, ϕ being a nondimensionalised circulation, and Re outlet Reynolds number. Also as ϕRe was reduced, the air core became thinner so in fact Woodford ignored the central lower w region in his subsequent analysis.

3.4. EFFECT OF VORTICES ON DISCHARGE COEFFICIENT C

The effect of vortices on the discharge is important and several authors have deduced empirical relationships between C and K. Daggett & Keulegan (1974) found, for Re $> 5\times10^4$:

$$C = 0.921 \qquad K < 0.1$$
$$C = 0.966 - 0.45\,K \qquad K > 0.1$$

where

K was defined as $\Gamma r_o/Q$ with $r = d/2$.

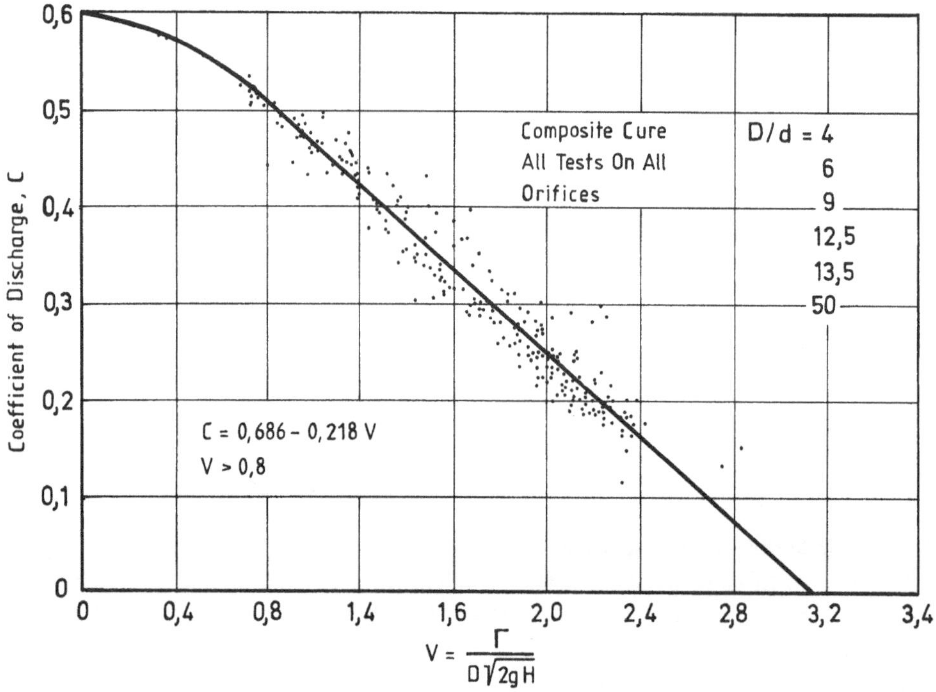

Figure 3.14. Coefficient of discharge versus Kolf vortex number.

The outlet was in the form of a rounded orifice. Figure 3.14 by Stevens & Kolf (1959) shows a similar linear dependence on K for $0.8 < K < \pi$, with no dependence on D/d in the range tested. K was defined as the Kolf vortex number with Γ/Vd, V being the vector resultant of u and v_t. The outlet was in the form of a sharp-edged orifice.

For $Re < 10^4$, Daggett & Keulegan (1974) found that viscous effects became evident, causing a decrease in discharge so that C became a function of only Re (Figure 3.15). Re was defined as $Q/r_o\nu$. Eventually the value of C coincides with that for pure radial flow towards the outlet, i.e. when $K = 0$. Weller (1975) points out that the radial Reynolds number $R_R = Q/\nu H$ is preferable to Re based on outlet diameter, since air-core penetration would change the effective value of this diameter. Furthermore, the above linear relationships predict negative values of C for large K, so there must be a limit to their validity. Anwar (1966) has investigated the behaviour of C as the circulation increase to higher values than observed by Daggett (Figure 3.16). The outlet pipe had a rounded entrance and protruded slightly from the tank floor. As the circulation constant $\Gamma r/Q$ (a form of Kolf number) increases, C decreases to an asymptotic value, rather than decreasing linearly to zero and/or negative values. Measurements were made well away from the air core which in these experiments was relatively slender.

Kamel (1964) obtained curves for discharge through nozzles of various sizes, and compared them with the inviscid theory of Binnie & Harris (1950) (Figure

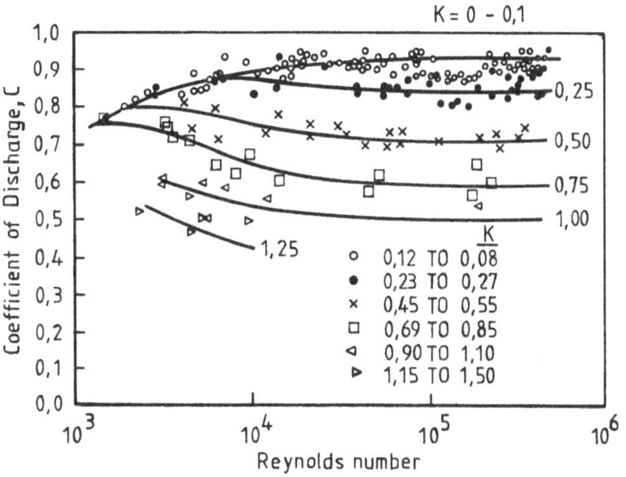

Figure 3.15. Effects of viscosity and circulation on coefficient of discharge.

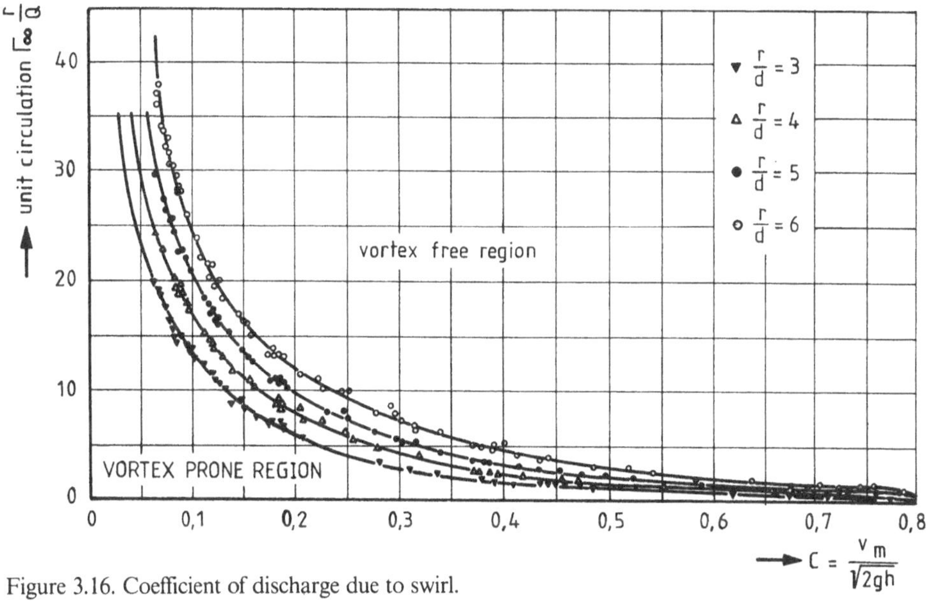

Figure 3.16. Coefficient of discharge due to swirl.

3.17). K in this case was defined as $\Gamma/r_0(2gH)^{1/2}$. On this theory, C should reach zero at K = 1 since the total static head should be completely used in giving tangential velocity to the fluid, thereby making the axial velocity zero. As can be seen this does not happen in practice, and Kamel suggests that this is due to boundary layer flow.

It seems that data are readily available for the variation of C with submergence or water depth H. Anwar & Amphlett (1985) have carried out experiments with a vertically inverted intake (i.e. flow upwards through a suction pipe) and found C to increase as H increased, for a given value of K. This trend is expected

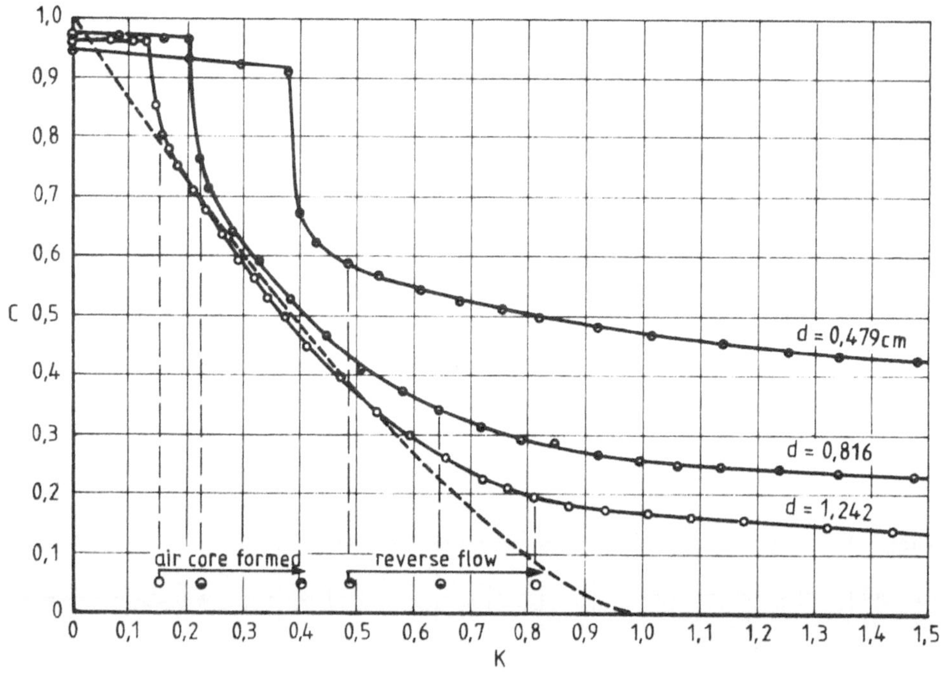

Figure 3.17. Coefficient of discharge for nozzles (broken line represents theoretical relationship of Binnie & Harris).

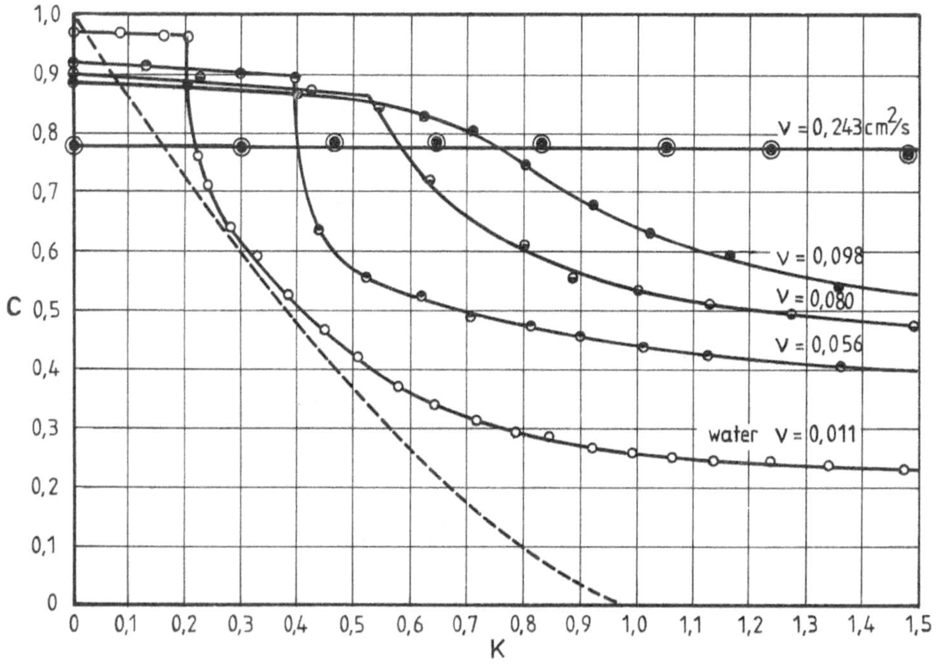

Figure 3.18. Coefficient of discharge for the 0.8 cm nozzle using oil.

to be similar for drain outlet also, since as H increases so the vortex becomes weaker and has less and less effect on the discharge.

3.5. EFFECT OF FLUID PROPERTIES

3.5.1. *Effect of viscosity*

This has been discussed indirectly in previous sections whenever the influence of Re or R_R has been mentioned. In general, as Reynolds number increases, so the effect of viscosity decreases, as expected, and the liquid behaves according to inviscid theory. Hence, above a certain critical value, the influence of Reynolds number can be ignored. Strictly speaking, the effect of viscosity can only be determined uniquely by using various liquids of different viscosities (or by testing one liquid at several temperatures). Kamel (1964) obtained the effect on C of increasing the viscosity by using different oils (Figure 3.18). For a given depth of liquid, higher degrees of swirl were needed to form an air core as the viscosity was increased.

As explained by Zielinski & Vilemonte (1968), as ν increases so Γ decreases more quickly from the outer boundary to the centre, hence the draw down in surface profile is less, i.e. the air core becomes narrower and weaker. This increases Q and hence C. Only when Re > 10^4 do the viscous effects become negligible and

$$C = f\left(K, \frac{D}{d}\right)$$

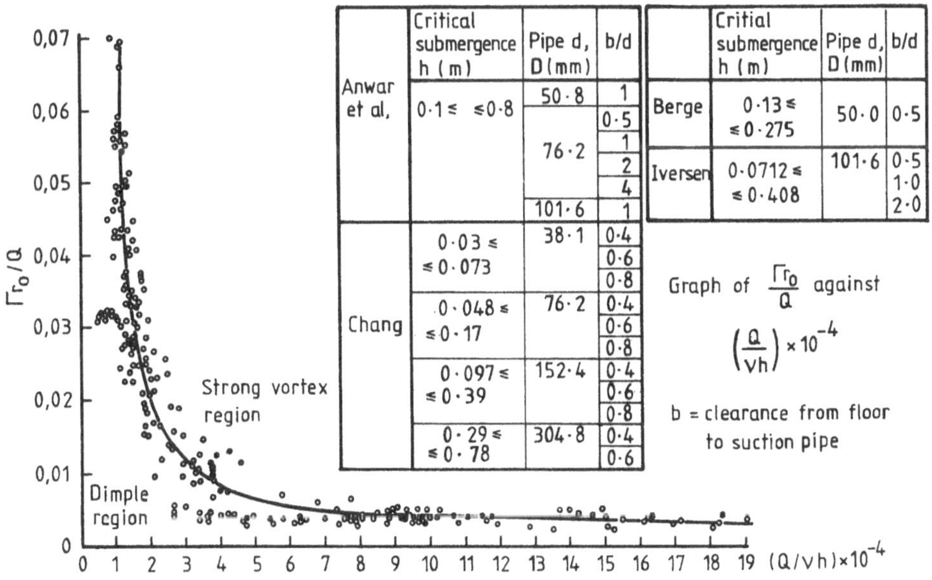

Figure 3.19. Effect of fluid viscosity on vortices forming at vertically inverted intake.

Anwar & Amphlett (1980) plotted results from various geometries of vertically inverted intakes, all using water to show that the effect of viscosity and hence radial Reynolds number, R_R, on $K = \Gamma r_o/Q$ was negligible for $R_R > 4 \times 10^4$ (Figure 3.19). This value will vary according to the intake geometry. For example, a value of $R_R > 10^3$ is given by Anwar, 1965 for vortices in a cylindrical tank with a bottom outlet.

Jain et al. (1978) used a chemical additive 'cepol' to increase the kinematic viscosity of water by a factor of 16.5 and found significant decreases in critical submergence with the more viscous solution.

The effect of viscosity is of major importance in predicting critical submergence, discussed later, and in deriving scaling laws to predict proto-type (real-life) performance from hydraulic models.

3.5.2. *Effect of surface tension*

The importance of this, and hence Weber number, on vortex flow has generally been considered to be negligible compared with gravitational and viscous forces, even for the case of a deep dimple vortex where surface tension would have its most pronounced effect because of the small radius of curvature of the surface at the vortex tip. Tests described below apply to the onset of air-entrainment, i.e. where the tip of the air core jut enters the outlet.

Daggett & Keulegan (1974) used fluids of the same ν but with surface tension differing by a factor 3 (0.072 to 0.024 N/m) and found no effect on the coefficient of discharge. Jain et al. (1978), using isoamyl alcohol added to water to decrease its surface tension from 0.072 N/m to 0.044 N/m, found no change in values of critical submergence within the range of Weber numbers tested, $11 < We < 184$. To put this range of We in perspective, for a typical model scale of 1 to 20, the value of We is decreased by a factor of 20. Hughes (1975) however, has suggested on theoretical grounds, that surface tension affects air entrainment in vortex flows. In another theoretical analysis, Yildirim & Jain (1981) found that surface tension produced a greater drawndown in the surface profile of a vortex (compared with a fluid with zero surface tension) when the imposed circulation was, $\Gamma^2/4\pi^2 g$, was less than about 2.8×10^{-5}. However, the theoretical analysis does not take account of viscous effects in the vortex core. There is, as yet, no experimental data to verify the theoretical analyses.

3.6. GENERAL CONSIDERATIONS ABOUT SIMILARITY CONDITIONS

The problem of scaling results from a physical model to predict full-size, prototype performance has been investigated intensively. It is apparent that one scaling law for one geometry will not necessarily be appropriate for another geometry. Thus, no definitive, universally applicable law can be expected to exist. Even for a given geometry, say the cylindrical tank under discussion in this review, one must be clear as to which parameter of the vortex flow is being modelled, e.g. C, h_{cr}, Γ, depth of dimple, volume of air entrainment, frequency or duration of air entrainment, formation conditions of a surface dimple, or the conditions for development of a dimple with tail into an air-cored vortex. The latter is, of course, the one of greatest practical importance in model studies.

From dimensional analysis, one obtains both geometric and kinematic dimensionless groups for the onset of vortex formation. These have been derived by several authors (Anwar & Amphlett, 1980; Jain et al., 1978; Daggett & Keulegan, 1974). The similitude requirements for the geometric dimensionless groups are met by ensuring geometric similarity on the model, i.e. an undistorted scale model. The remaining dimensionless groups are Fr, Re, We and K, K being the Kolf number, which has been modified and called a circulation number or Rossby number by some authors. It is impossible to meet the similitude requirements of these four parameters if water is used in both model and prototype, even if We is taken as insignificant. Various authors have suggested scaling laws to overcome this difficulty, based on theoretical, experimental or prototype investigations.

It is important to stress that no single simple scaling law can be expected to apply over the whole range of geometries and types of vortex. For the initial formation of a free surface depression Berge (1966) states that Fr is important. This is confirmed by Chang (1982) who gives some data to support this for the case of a rectangular geometry, and by Hecker (1981) in a comprehensive comparison of model and prototype data.

However, for the case of an air entraining vortex, where the air core is just about to enter or has already entered the outlet, it is generally agreed that scale effects do exist. One exception is Quick (1962) who found C to depend on K for a strong stable vortex. He argued that as K was a form of Froude number, so Froude scaling should be used for modelling discharge. Hecker, in model-prototype comparisons, found that scale effects existed on Froude scale models when simulating air core vortices. To overcome the scale effects, several approaches have, in general, been adopted: (a) to determine the Reynolds number Re or R_R above which viscous influences are negligible. Above this Reynolds number the correct circulation patterns will be produced if the approach geometry is correctly modelled, and the model can then be run according to Froude scale; (b) to determine a correction or multiplying factor to allow for viscous effects. The factor is applied either to the results of a model run at Froude scale or to the model velocity so that the model is ran at higher than Froude scale; (c) to operate the model with increased water temperatures so as to be closer to prototype Reynolds numbers. These approaches are now considered in turn.

For vortices with a narrow air core and for strong steady vortices Anwar (1965, 1966) found that viscous influence could be neglected for $R_R > 10^3$. Furthermore, he suggested a model scale ratio of not less than 1 to 20 for a Froude scale model to reproduce prototype vortex formation. He also pointed out that excessive surface roughness on the tank floor reduced tangential velocities and could even suppress vortex formation. Also, although the onset of air entraining vortices could be correctly modelled the actual size of the air core is smaller on a model and is not correctly scaled. In a later study, Anwar & Amphlett (1980) suggested a figure of $R_R > 4\times10^4$ for a different geometry, that of a vertically inverted intake subjected to a superimposed circulation. For yet another geometry that of a closed conduit spillway, Donnelly et al. (1974) show that the entrance loss coefficient is constant for $Re > 1.5\times10^5$. On the other hand, Daggett, using a bottom outlet, obtained a limiting value of about $Re > 3\times10^4$ above which viscous effects on C, and hence presumably vortex formation, were negligible. Jain concluded that the Reynolds number at which viscous effects become negligible is dependent on Froude number. He expressed his result in the form of a dimensionless number N_ν which is effectively the ratio of Re/Fr. The limiting value is $N_\nu > 5\times10^4$. This is equivalent to $Re > 5\times10^4 Fr$, both Re and Fr being based on outlet diameter. The numerical value is however questioned by Amphlett (1979). Furthermore, Dhillon et al. (1981) using various models built and operated according to the criteria above, found that Jain's criteria was not correct for their geometry, and that Anwar's (1965) gave the best correlation with observed prototype performance. Nystrom et al. (1982) in describing tests for model studies with scale ratios typically between 1:2 and 1:4 of sumps which have unusual geometries, suggests the use of $Re > 7\times10^4$ and $R_R < 1.5\times10^4$ for modelling free surface vortex strength, persistence and consequent air entrainment. Also, flow swirl angles were independent of scale ratio when the approach Reynolds number $R_a > 3\times10^4$, based on the velocity and depth at the containment floor. These results are also based on comparisons with fullscale performance.

The second general approach of applying a correction or multiplying factor has been suggested by several authors. Jain et al. (1978) have suggested applying a correction factor, greater than 1, to the results obtained for critical submergence on a model operated at Froude scale. Whilst this has the advantage of no velocity and hence no surface turbulence exaggeration, it does assume that approach flow patterns are independent of water depth, which is not always true for irregular geometries.

Another form of correction has been suggested by Knauss (1972). The model is still run according to Froude scale but the outlet (or intake) cross-section is distorted to give greater than Froude scale exit velocities. This method is only applicable for modelling air entraining vortices, and is not necessary when only investigating surface swirling.

The use of a multiplying factor, applied to the Froude scale model velocity, has been used for some time now for various vortex studies. However, there has not been general agreement on actual numerical values to use. Only two systematic studies have been made along these lines (Zajdlik, 1977; Chang, 1980a) and both were for rectangular sump geometries, so the results would only apply qualitatively to cylindrical geometries. As expected, the multiplying factor was found to increase as the model scale ratio, and hence Reynolds number decreased. Also, the multiplying factor was found to increase with submergence, a factor previously ignored.

A novel 'vortex projection' method making use of the decrease in ν for water as temperature is increased, was developed by Durgin & Hecker (1978). This technique has been applied mainly to model studies of reactor sumps and has the advantage of no need for excessive velocity exaggeration which could distort flow patterns. Where model Reynolds numbers are already above the limiting values discussed above, there should be no need to use this technique. No change in vortex strength of persistance with temperature was found by Padmanabhan & Hecker (1982) when the model Re was high enough.

Only one author, Haindl (1959), has found that Reynolds number scaling was appropriate.

Various theoretical analyses have been presented, and provide support for increasing model velocities above Froude scale. Polikovski & Perelman (1956) suggested a local scaling of the vortex core according to Re, and Fr scaling for the outer flow. Holtorff (1964) also derived similar results theoretically, dividing the flow into two regimes, $r > r_0$ where Fr scaling applies and $r < r_0$ where both Fr and Re must be scaled. This however is impossible to achieve in practice. Rohan (1966) discussed the effect of Fr scaling on viscous shear forces, and showed that tangential stresses on the model would be too large on the model, thus reducing angular velocities and also the likelihood of vortex formation. Therefore, he suggested that model velocities should be increased above Froude scale by an amount depending on the ratio of prototype and model Reynolds numbers.

The 'equal model-prototye velocity' rule is mentioned here for completeness. Kenn & Zanker (1967) explains theoretically how such a rule is justified if a dimensionless capillary number, derived from the ratio of We/Re, is considered to be the most important parameter.

A further parameter not often considered is correct modelling of the vorticity flux approaching the intake. Chang (1980b), Hattersley (1965), Hecker (1981) and Quick (1970), mention this parameter, and scaling requires careful consideration of boundary roughness and Reynolds number related to the vorticity source (Farell & Cuomo, 1982). This is probably a secondary consideration for drain outlets but Chang (1980) has suggested that for vertically inverted intakes, the intake pipe itself is the major source of vorticity. Spurr (1980) emphasises the importance of maintaining correct turbulence levels in a model, since turbulence can aid in diffusing vorticity or hinder the organisation of vorticity into a coherent vortex. However, it is not clear how this vorticity scaling can be achieved in practice.

Prediction of critical submergence 4

J.KNAUSS
Technical University of Munich, Obernach, FR Germany

4.1. INTRODUCTION

In order to avoid air entrainment by free surface vortices and to minimize swirl entrainment by subsurface vortices an intake must be adequately submerged. Therefore, one of the most important considerations for the design of intakes is how to predict the necessary depth to which they should be submerged. Presently, there exist no completely satisfactory answer to this question. In this chapter of the monograph a method is described to achieve a more accurate solution of the hydraulic problem involving two special aspects in the evaluation of the results of some recent studies of intake performance.

The first special aspect of this method is the use of a combined dimensionless number which would improve the description of the influence of induced circulation on the critical conditions for the hydraulic performance of intakes. This aspect enables a simplified and unified representation of the different test results as well as a convenient comparison of them. Further, it provides an improved insight into the hydraulics of intakes.

The second aspect involves consideration of the role which seems to play the intake orientation on the prediction of the critical submergence for the intake. This is done by considering two characteristic dimensions of the vortex shape for the different intake types. A certain influence of the intake direction on submergence requirements is apparent comparing the results of the various experiments.

Submergence requirements were determined by definition and observation of the critical intake stage (see Chapters 1 and 2). In this evaluation, the following criterion was used: Production of circulation in the approach flow and concentration of swirl above the intake just reaches an amount which enables a free surface vortex to penetrate the submergence depth. From the tail of the open air-core bubbles are sucked into the intake. The air entrainment is intermittent but continuous. The entrained air volume is of a measurable amount (see vortex Type 5 after Durgin & Hecker, 1978).

Figure 1 gives the chosen definition of the intake direction ($\vec{\phi}$) and the submergence

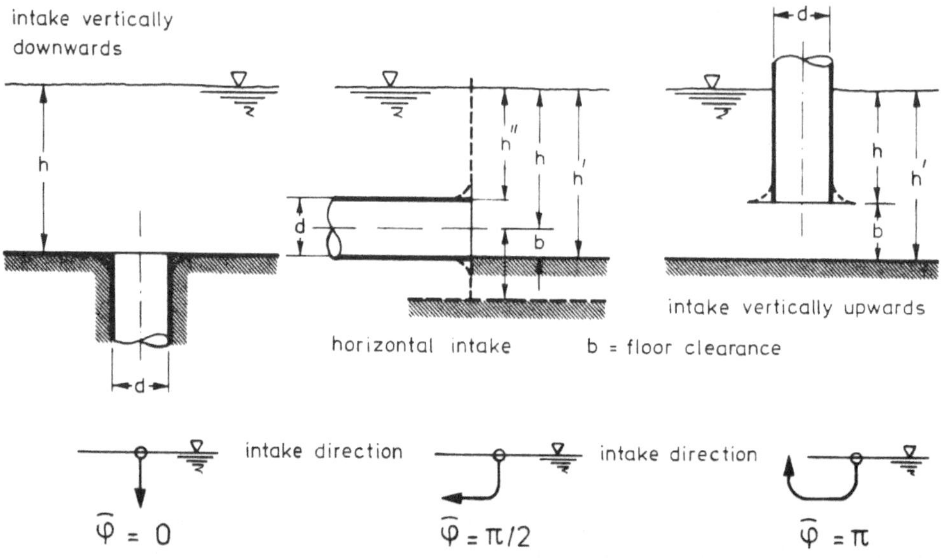

Figure 4.1. Definition of intake direction ($\bar{\phi}$) and submergence depth (h).

depth (h). Combination of h with d (diameter of the circular intake, or a typical linear dimension of a non-circular entrance section) gives the relative intake submergence (h/d).

Establishing characteristic dimensionless numbers relevant to the hydraulic system under consideration (Fr, N_c, Re, We) in this chapter of the monograph the diameter d of the intake is introduced as the governing geometrical dimension. For dimensional analysis the intake rate Q was chosen as a second independent variable of the system. Both parameters are typical and most important design values. Together they define the intake velocity ($v = 4Q/d^2\pi$).

The following dimensionless numbers are introduced:

$$Fr = \frac{v}{\sqrt{g\,d}} \qquad \text{(intake Froude number)}$$

$$Re = \frac{v\,d}{\nu} \qquad \text{(intake Reynolds number)}$$

$$We = v\sqrt{\frac{\rho\,d}{\sigma}} \qquad \text{(intake Weber number)}$$

Moreover, swirling flow or vortex flow at intakes requires the introduction of a further dimensionless number describing the influences of circulation on the flow conditions (see Chapter 2). The 'circulation number' used in this investigation is given by the term:

$$N_c = \frac{c}{v\,d}$$

where

$$c = v_t\, r \quad \text{(circulation constant).}$$

The circulation number N_c is a particularly important number of the hydraulic system because it relates a governing hydraulic parameter (c) to the main design values of the intake (Q or v and d). The circulation constant can be regarded as the hydraulic reaction on the approach flow properties and the location of the intake relative to its surroundings (open channel or basin). The strength of circulation is determined by the upstream boundary conditions and is dependent on the geometry of the approach flow area.

In order to determine the relative critical submergence for an intake the following equation, which involves all pertinent parameters, can be formulated:

$$(h/d)_{cr} = f(\overline{\varphi}, Fr, N_c, Re, We)$$

Under certain conditions it is possible to reduce the series of dimensionless numbers influencing the critical intake submergence (see Chapters 2 and 3). The intake Weber number, representing the effects of surface tension, is assumed not to be important at the critical stage of vortex formation at intakes. The role of fluid viscosity may be neglected if the intake Reynolds number exceeds a certain limit ($Re > 10^5$). The equation consequently reduces to

$$(h/d)_{cr} = f(\overline{\varphi}, Fr, N_c) \tag{4.1}$$

A further simplification is available if one makes use of a suitable combination of the remaining dimensionless numbers. An analysis of the test results from a series of recent studies of the hydraulics of intakes (see the details in Knauss, 1983) suggests that the circulation number (N_c) can be appropriately combined with the intake Froude number (Fr), to yield

$$N_c\, Fr = \frac{c}{\sqrt{g}\, d^{\frac{3}{2}}} \tag{4.2}$$

The product may be called the 'swirl parameter' which includes both the geometric and hydraulic characteristics of the approach flow and of the intake for a circulating inflow. The critical submergence equation may now be cited as

$$(h/d)_{cr} = f\left(\overline{\varphi}, \frac{c}{\sqrt{g}\, d^{\frac{3}{2}}}\right) \tag{4.3}$$

with the intake direction as an independent parameter of the hydraulic system (Figure 4.2).

The application of the derived swirl parameter is directly dependent on the exact estimation of the circulation constant c and on the relation of the determining value of c to the critical intake conditions. In Figure 4.3, a schematic representation of the complete hydraulic system for an intake is shown. It illustrates the important distinction between the approaching swirl (c_a), the critical swirl (c) and the entrained swirl (c_e).

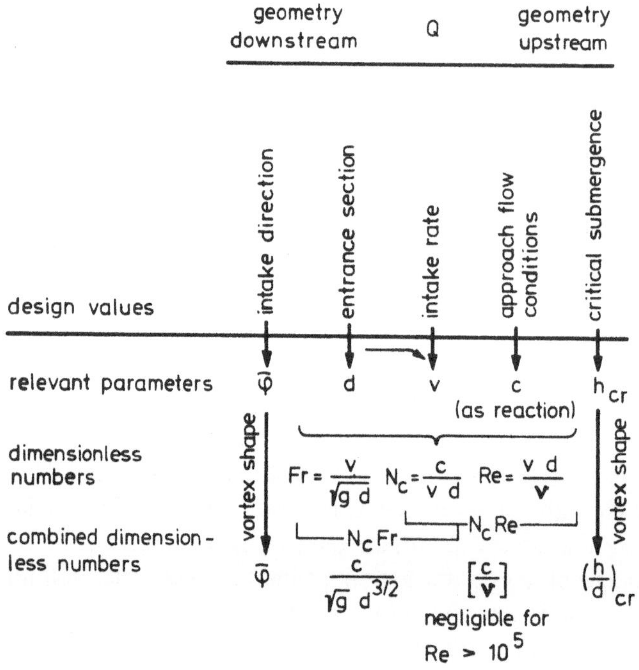

Figure 4.2. Development of governing flow parameters.

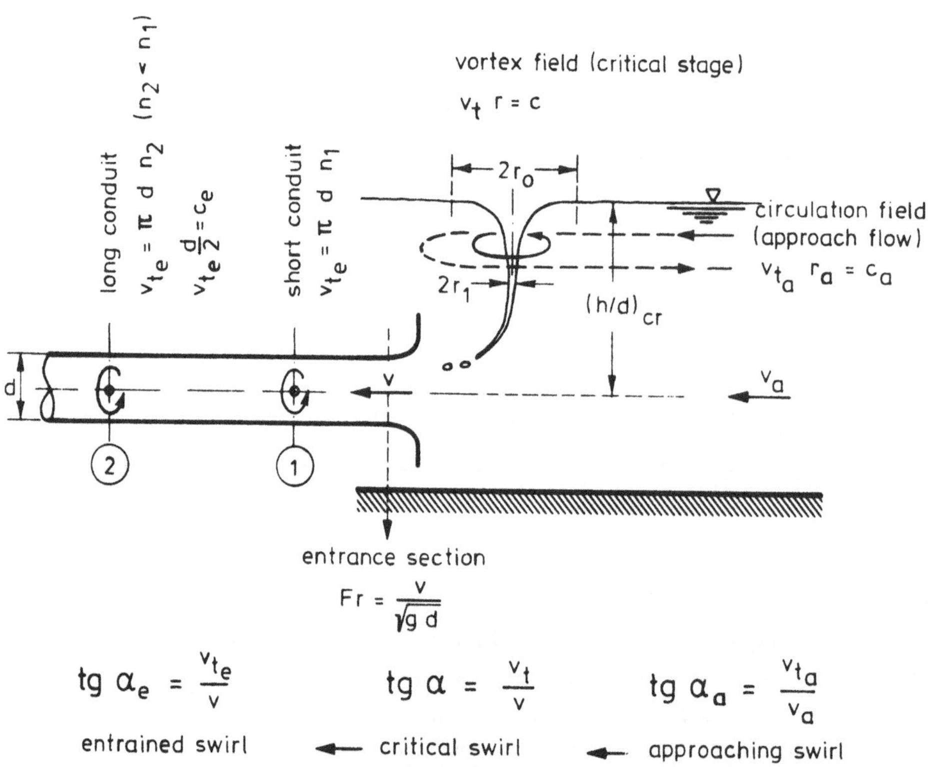

Figure 4.3. Schematic representation of the complete hydraulic system for an intake.

A complete investigation involving measurement and correlation of:

(c_a with c) and (c with c_e) with respect to $(h/d)_{cr}$

has not yet been carried out. Until now, information is only available from test procedures correlating parts of the complete system.

The determination and the use of the circulation constant as the main parameter governing the flow pattern under discussion is one of the major problems requiring further research for intakes (see Chapter 2).

4.2. REVIEW OF RESEARCH RESULTS

4.2.1. *Critical submergence versus special selected parameters*

In order to avoid the occurrence of air entraining vortices at intakes of power plants, in former times the calculation of minimum submergence was related to the velocity head at the entrance section (similar to the estimation of inflow losses):

$$\text{required minimum } h = \text{const.}\ \frac{v^2}{2g}$$

where

$$\text{const.} \geqq 2.$$

For intakes in rectangular pump sumps Prosser (1977) recommends a very simple relationship for the necessary minimum submergence:

$$\text{required minimum } h = \text{const. } d$$

where

$$\text{const.} \geqq 2.5.$$

From Prosser's assumption of a mean intake velocity of v = 4.0 m/s (which he considers as typical for a modern axial flow pump), one can derive the order of the intake Froude and the intake Reynolds number which are commonly used in design practice (especially in pump sumps):

$$0.5 < Fr < 2 \text{ and } 10^6 < Re < 10^7 \text{ for } 1 < Q < 100\ m^3/s$$

The recommendations of Prosser have to be restricted to small - or medium - size installations. For instance, a minimum submergence depth of more than twice the intake diameter may not be an economic solution in the case of a large pumped storage scheme because excessively large dimensions may result for the basin.

For large size installations, a mean intake velocity of v = 2.0 m/s is usual. Therefore, the resulting range of the dimensionless numbers is indicated by:

$$Fr < 0.5 \text{ and } Re > 10^7.$$

In some investigations (e.g. Chang, 1979b; Dicmas, 1978) the relative critical

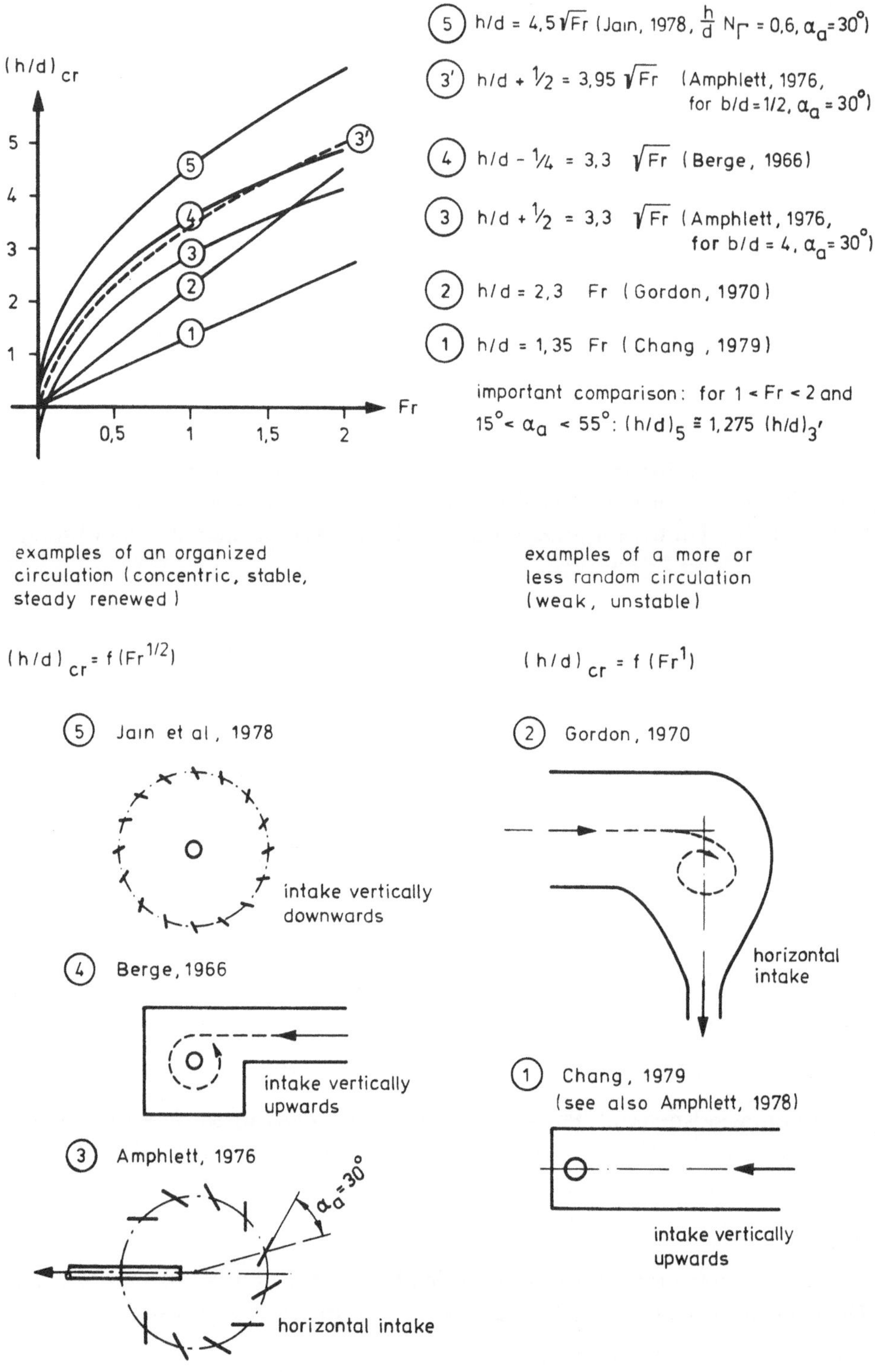

Figure 4.4. Review and comparison of $(h/d)_{cr} = f(Fr)$.

submergence of an intake is directly related to the intake velocity:

$$(h/d)_{cr} = f(v)$$

The results leading to this kind of equation are strongly related to the geometrical configuration of the investigated intake system and may only be generalized within limited boundary conditions.

An often used method is to evaluate the results from tests with special geometrical arrangements by relating the relative critical submergence to the intake Froude number. The results are usually presented in the following form:

$$(h/d)_{cr} = \text{const.}\ Fr^{x}$$

In literature different values for the exponent x were suggested. For example, $x = 2$ (Quick, 1970), $x = 1$ (Gordon, 1970; Reddy & Pickford, 1972) and $x = 1/2$ (Jain et al., 1978). Only one experimental confirmation was found for the equation with $x = 2$ (Humphreys et al., 1970) but for $x = 1$ and $x = 1/2$ a series of experimental results are available as shown in Figure 4.4.

A significant difference between the tests with an organized and stronger circulation ($x = 1/2$) and those with a random and weaker circulation ($x = 1$) is evident. The minimum value of the proportional factor of the discussed relationship was found at least to be in the order of const. = 1. The factor increases as the circulation becomes stronger. In all experiments two-dimensional approach flow conditions were installed using vertical walls in the surrounding basin or channel.

From his extensive observations at prototypes, Gordon in 1970 derives a linear limit, $(h/d)_{cr}$ = const. Fr, between intakes with and without vortex problems. He makes a general but not well defined distinction between symmetrical (const. = 1.7) and asymmetrical (const. = 2.3) approach flow conditions. The proportional factor increases with increasing circulation. The observed value of the two constants

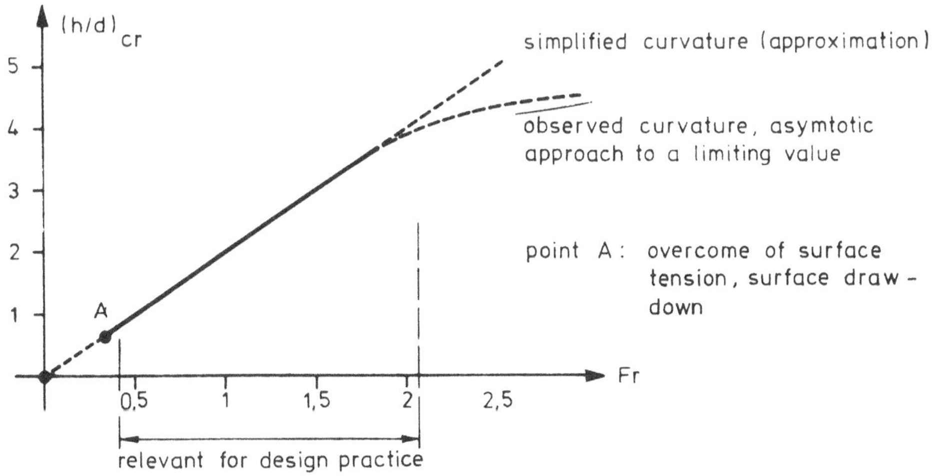

Figure 4.5. $(h/d)_{cr}$ = K Fr within a limited range of intake Froude numbers.

seems to be a result governed by chance as one can see from a comparison of Gordon's investigations with the results of the tests with an exactly defined geometry (Figure 4.4).

The most important outcome of Gordon's prototype investigation is that for the purpose of design the relation between the critical submergence and the intake Froude number may be taken as a linear function. In the range of practical Froude numbers (see Figure 4.5) and for a normally given 'moderate' vortex formation, one can assume

$$(h/d)_{cr} = K\ Fr \tag{4.4}$$

with the proportional factor K generally depending on the geometrical conditions and the strength of the initial circulation.

4.2.2. *Critical submergence versus all governing parameters*

The first test evaluating the dependence of the critical submergence on the induced circulation was carried out and published by Berge in 1966. He investigated an intake directed vertically upwards and situated in a special invariable and asymmetrical approach flow geometry (see Figure 4.4). Berge developed the so-called 'refracted ray method' for measurement and determination of the critical swirl (see Chapter 5).

Until now, the results of four extensive investigations on the hydraulic performance of intakes with respect to the induced circulation are available. In all test arrangements variable approach flow conditions could be adjusted in order to alter the initial circulation over a wide range. The details of the different tests (set up, measuring technique and outcome) are given in length in the original publications or in a summary concentrated on especially interesting aspects published by Knauss in 1983. For further evaluation, discussion and comparison of the various results, it is very important to notice the different methods of measuring the occurring circulation and also the definition of the critical intake stage. The four investigations mentioned here contain the three main intake directions (see Figure 4.1). The influences of fluid viscosity and surface tension were examined. Certain limits were developed defining when their influences may be neglected (see Chapters 3 and 5).

After a transformation of the individual variables, used for the analytical interpretation of the various test results, into the defined characteristic parameters of this evaluation, one gets the following series of equations which describe the observed relation between the relative critical submergence, the swirl parameter and the type of intake investigated:

Daggett & Keulegan (1974): vortex orifice flow

$$(h/d)_{cr} = \frac{300}{Fr} \frac{c}{\sqrt{g}\, d^{\frac{3}{2}}}$$

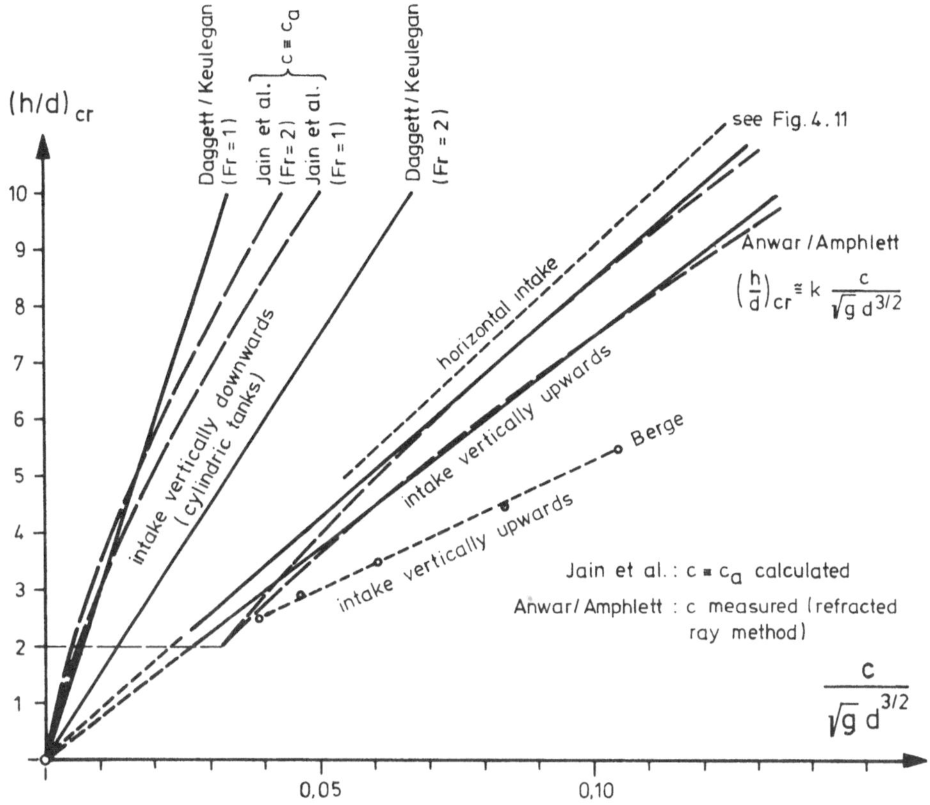

Figure 4.6. Review and comparison of $(h/d)_{cr}$ = f (swirl parameter and intake type).

Jain et al. (1978): intake vertically downwards*

$$(h/d)_{cr} = 88\ Fr^{0.139}\left(\frac{c}{\sqrt{g}\ d^{\frac{3}{2}}}\right)^{0.724}$$

Anwar et al. (1978): horizontal intake

$$\frac{0.86\ (0.054\ (h/d)_{cr} + 0.139)^2}{\left((h/d)_{cr} + 0.5\right)^{\frac{1}{2}}} = \frac{c}{\sqrt{g}\ d^{\frac{3}{2}}}$$

Anwar & Amphlett (1980): intake vertically upwards

$$\frac{0.86\ (0.060\ (h/d)_{cr} + 0.126)^2}{\left((h/d)_{cr} + 0.5\right)^{\frac{1}{2}}} = \frac{c}{\sqrt{g}\ d^{\frac{3}{2}}}$$

*With c calculated from the approach flow angle α_a, therefore c used in the equation represents c_a (see Figure 4.3).

These equations are evaluated and plotted in Figure 4.6. The character of the resulting lines and curves enables the following statement to be made: An approximately linear relationship between the critical submergence and the swirl parameter (combined dimensionless number, see Section 4.1) seems to be characteristic for all types of intakes. Such a simplification of the test results may be permissible taking into account the difficulties and uncertainties of the necessary measurements or calculations (governing circulation constant) and of the special observations (recognition of critical intake stage). The simplified equation reads

$$(h/d)_{cr} = k \frac{C}{\sqrt{g}\, d^{\frac{3}{2}}} = k\, N_c\, Fr \qquad (4.5)$$

with a proportional factor k representing the gradient of the linear relationship (see also Figure 4.11).

The comparison of the different lines in Figure 4.6 yields a second statement involving the physical meaning of the factor k: For the same value of the swirl parameter, the submergence requirements are different due to the intake type. Or, with an equal submergence the critical circulation which produces an air entraining vortex changes with the intake type. Therefore, the factor k in the above equation may represent the influence of the intake direction.

4.3. ROLE OF INTAKE DIRECTION ON CRITICAL SUBMERGENCE

The vortex shape expressed by the length of the vortex tube and the thickness of the open vortex core seems to be an important indicator of the influence of intake direction on the critical conditions as schematically shown in Figure 4.7 (see for example the investigations of Amphlett, 1976, 1978). Passing from the left to the right side in Figure 4.7 one can see an increase of the length and the thickness of the vortex air-core. For the depicted constant submergence depth,

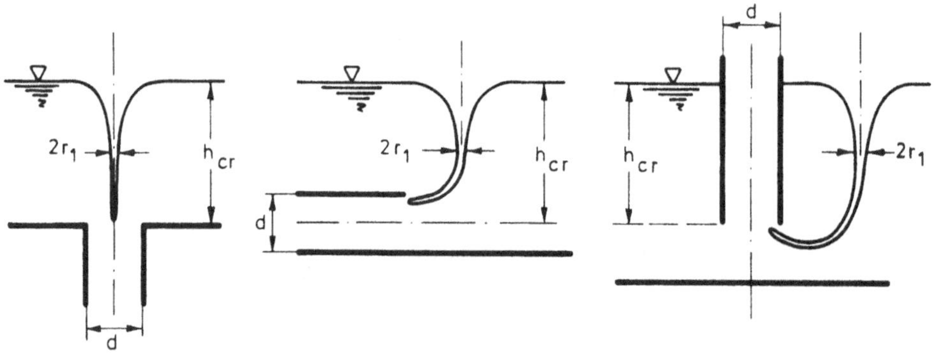

Figure 4.7. Vortex shape versus intake direction.

the critical circulation also increases. From the results of research it is known (see Chapter 3) that for a longer combined vortex (line type) the open throat must be wider than for a shorter one. Therefore, the air-core radius (r_1) at a certain elevation above the intake should be a characteristic proportion determining the vortex shape relative to the intake type.

The prediction of the core radius and its relation to the tube length regarding the different intake directions is the main scope of the next section.

4.4. EVALUATION OF SPECIAL TEST RESULTS

4.4.1. *General relationship between vortex shape and swirl parameter*

A theoretical possibility to express the relationship between vortex shape and swirl parameter offers the conception of Rankine for the so-called 'combined' vortex in an ideal, non-viscous fluid. The term 'combined' stands for the hypotheses that a line vortex or dimple may be composed of a rotating core and a circulating field around the core. The boundary between the two ranges is marked by the width of the air-core ($2r_1$) at the defined position shown in Figure 4.8. This pattern results from the assumption of a simplified distribution of tangential velocity (v_t) over the whole range (see Chapter 2). The simplification means that one considers a 'virtual' throat diameter ($2r_1$) but not a real value.

The depth of the combined vortex equates to

$$h = \frac{c^2}{g\, r_1^2} \text{ or } \frac{h}{d} = \frac{c^2}{g\, d\, r_1^2}$$

if one relates the vortex to the intake. A rearrangement of the last equation yields

$$\frac{r_1}{d}\sqrt{\frac{h}{d}} = \frac{c}{\sqrt{g}\, d^{\frac{3}{2}}} \tag{4.6}$$

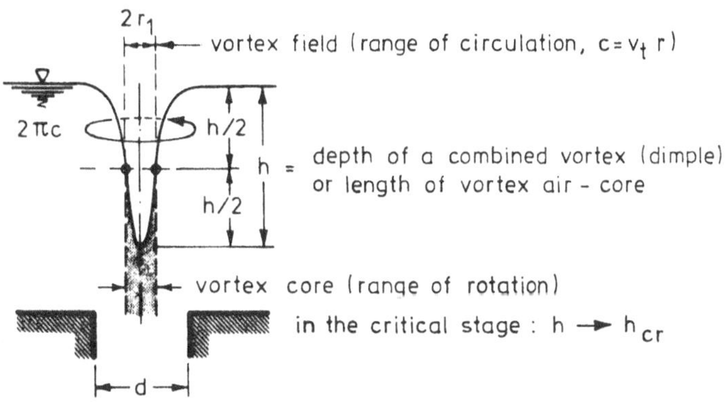

Figure 4.8. Combined vortex (Rankine vortex).

where

r_l/d_l = relative width of the vortex air-core
h/d = relative length of the vortex air-core.

Equation (4.6) demonstrates the relation between two typical dimensions of the vortex shape and the swirl parameter for the occurrence of a dimple (see Knauss, 1983) and also for the critical intake conditions. The combination of Equations (4.5) and (4.6) expresses

$$(r_l/d)_{cr} = \frac{1}{k}\,(h/d)_{cr}^{\frac{1}{2}} \tag{4.7}$$

where

$k = f(\bar{\varphi})$ as assumed above (see Section 4.2.2).

Equation (4.7) means, that a linear relationship between the critical submergence and the swirl parameter is valid, if the relative width of the vortex air-core is directly proportional to the square root of the relative length of the vortex tube. If one can measure or calculate (r_l/d) and its relation to (h/d) for the critical intake conditions at different intake types, then the gradient k in Equation (4.5) can be evaluated.

4.4.2. *Air-core width and length relations for surface vortices at critical intake conditions*

Figure 4.9 shows a very interesting test result from Amphlett (1976) obtained in his investigation of a horizontal intake.

The diagram enables an evaluation of (r_l/d) versus (h/d) as pointed out in the figure. The result is given in Figure 4.10 which allows the following conclusions: For each line $(h/d)_{cr}$ the virtual relative air-core width (r_l/d) is nearly constant. The small scattering of the measured values enables the introduction of a mean value $(r_l/d)_m$ and its relation to the square root of (h/d). The interesting outcome of this comparison is a straight line which confirms the expected linear relationship between the two vortex shape parameters. The line does not meet the zero-point of the diagram which is physically necessary. An explanation of this fact may be obtained from a view on the test set-up. For production of different circulations a vane assembly was used which fixes the center of the formed vortices at a certain distance ($\sim$ 1.5 d) in front of the intake (see Amphlett, 1976). If one adds this constant horizontal distance to the variable vertical distances (h/d) in order to get the total length of the vortex tube, then the straight line starts with zero for both parameters. Equation (b) in Figure 4.10 gives the analytical expression of the investigated relationship.

The resultant gradient of the line is assumed to be a characteristic value for a horizontally directed intake. The proportional factor in Equation (4.7) could now be determined as

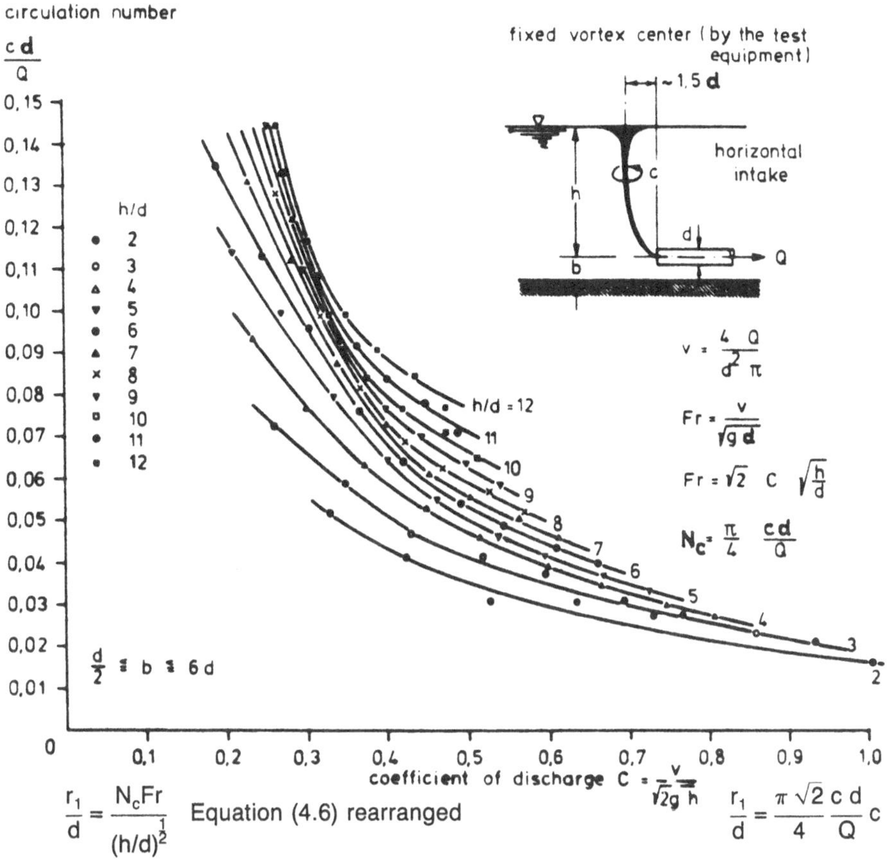

Figure 4.9. 'Effect of circulation on coefficient of discharge for the onset of air-entraining vortices' (Amphlett, 1976; Anwar et al., 1978).

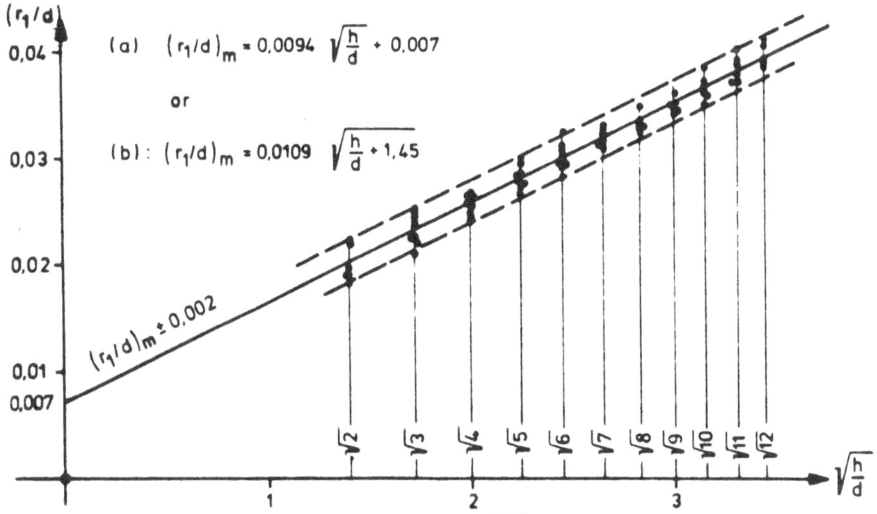

Figure 4.10. Result of evaluation $(r_1/d) = f(\sqrt{h/d})$ for critical intake conditions.

$$\frac{1}{k} \cong 0.011 \quad \text{(horizontal intake)}$$

The corresponding value for an intake inverted vertically downwards can be obtained rearranging the critical submergence equation developed by Jain et al. (1978) and comparing it with Equation (4.6). Refering to the detailed description in Knauss (1983) the final result of this evaluation is

$$\frac{1}{k} = 0.004 \frac{c}{c_a}$$

From special test results of Amphlett (1976) one can derive a mean order of $c/c_a = 2.2$ (relation of critical swirl constant to approaching swirl constant, see Figure 4.3) valid for a test arrangement using a guide vane assembly for variable swirl production (Knauss, 1983). Therefore, the relevant proportional factor of Equation (4.7) in this case must be corrected to

$$\frac{1}{k} \cong 0.009 \quad \text{(intake vertically downwards)}$$

As already shown in Figure 4.6 the gradient k for an intake vertically downwards is steeper than the comparable value for a horizontal intake.

4.4.3. *Critical submergence versus swirl parameter*

Amphlett (1978) gives the results of his investigation of the horizontal and the vertically upwards inverted intake in the way of $h_{cr} = f(c)$. He defines a linear relationship between h_{cr} and c. After a transformation of $h_{cr} = f(c)$ into $(h/d)_{cr} = f\,(c/\sqrt{g}\ d^{2/3})$ the resulting plots allow a direct evaluation of the gradient k of the obtained straight boundary line for the critical intake conditions. Figure 4.11 shows two examples of such plots. Further examples are given by Knauss (1983).

As one can see, the experiments yield different submergence depths at the same swirl parameter. This fact must be related to the special nature of vortex formation and to the problems of swirl measurement as well as to the difficulties with the exact determination of the critical stage when air entrainment starts (see Chapter 2). Therefore, it is suitable to place the boundary line at the upper range of the observed scattering values. This procedure enables a safe prediction of required submergence applying the experimental results. If different critical submergences are possible with the same approach flow conditions, then the larger one has to be taken as the design criterion.

From Figure 4.11 one can obtain at first a confirmation of the previously found gradient k for the horizontal intake and, secondly, for the sake of completion, the vortex shape parameter of the intake directed vertically upwards:

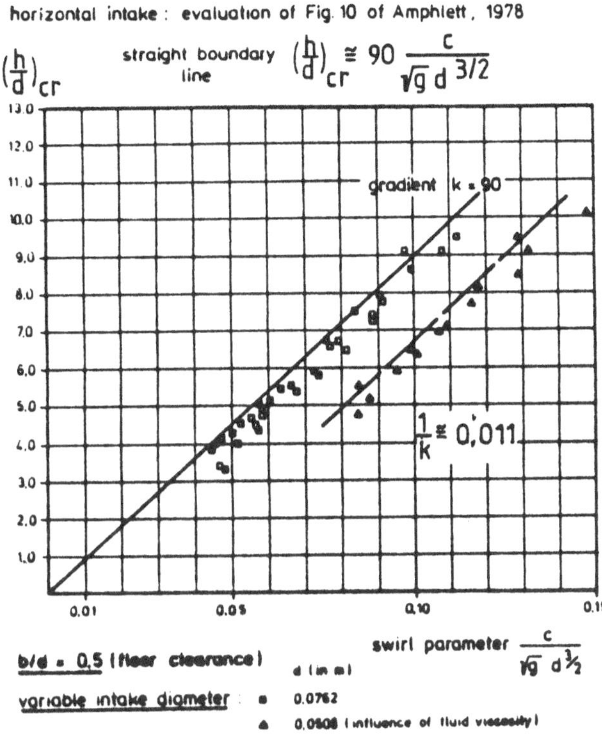

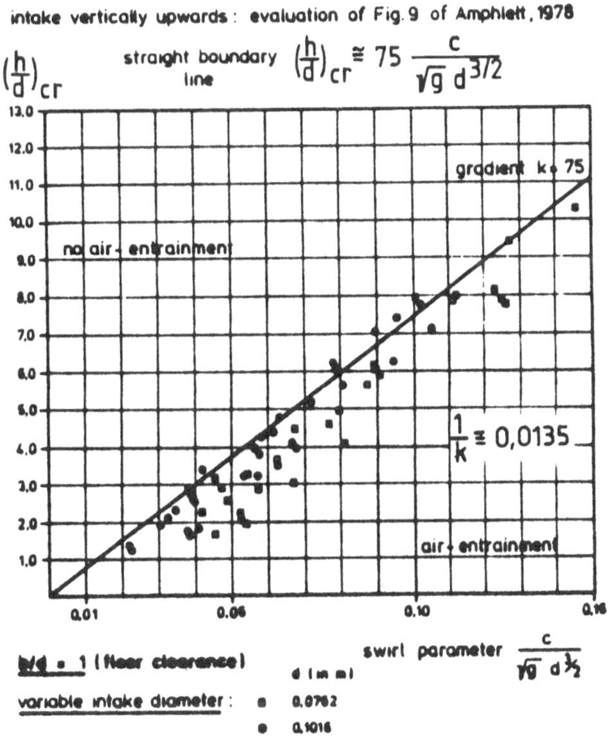

Figure 4.11. Direct evaluation of gradient k.

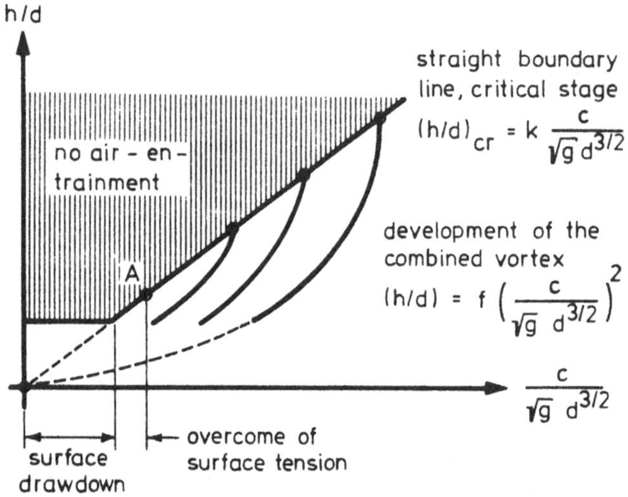

Figure 4.12. Details of the boundary line for critical conditions.

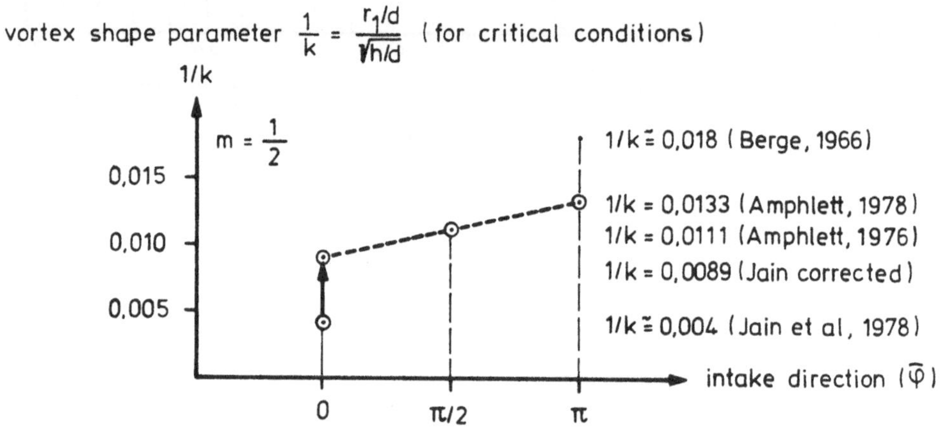

Figure 4.13. Vortex shape parameter versus intake direction.

$$\frac{1}{k} \cong 0.0135 \quad \text{(intake vertically upwards)}$$

In Figure 4.12 some details of the boundary line for the critical conditions are given, especially a comparison between the curved line of dimple development and the approximated straight limit for incipient air entrainment.

4.4.4. *Final results*

In Figure 4.13 all the collected data concerning the dependence of the vortex shape parameter (1/k) on the intake direction ($\bar{\phi}$) are plotted. A further linear approximation of the resulting relationship can be formulated

$$\frac{1}{k} = \left(\frac{1}{k}\right)_{\bar{\phi}=0} \quad \left(1 + m\,\frac{\bar{\varphi}}{\pi}\right) \cong 0.009 \left(1 + \frac{\bar{\varphi}}{2\pi}\right) \tag{4.8}$$

which offers an analytical approach to the necessary information about the gradient k in Equation (4.5).

One advantage of this formulation may be the possibility of an interpolation of k-values for inclined intakes (see Chapter 1). The second one is the simple production of relative k-values enabling the transformation of the experimental results from an examined intake to a not yet investigated type (for example: $k_o/k_{\pi/2} \cong 1.25$ and $k_o/k_\pi \cong 1.5$).

Moreover, the knowledge of the factor k, supposed to represent the influence of intake direction on the critical conditions, enables the establishment of a numerical relationship between the two main equations for the relative critical submergence depth, $(h/d)_{cr}$ = K Fr (Equation (4.4)) and $(h/d)_{cr}$ = k N_c Fr (Equation (4.5)), which finally determines the critical circulation number crit. N_c = K/k as the ratio of the two proportional factors. If K is known for example from prototype observations as given by Gordon (1970) (e.g. K = 2.3 for asymmetrical approach-flow conditions at a horizontally directed intake) and k is read from Figure 4.13 with $k \cong 90$ for $\bar{\phi} = \pi/2$ then the critical circulation number is of the order of crit. $N_c = 0.025$. Refering to Figure 4.3, in the vortex field the critical swirl angle is defined as $tg\alpha_{cr} = v_t/v$. The critical circulation number generally reads crit. $N_c = tg\alpha_{cr}$ (r/d). Related to the intake wall (r/d = 1/2) one gets crit. $N_c = tg\alpha_{cr}/2$. With crit. $N_c = 0.025$ the critical swirl angle at this location yields $\alpha_{cr} \cong 2.9°$ for a horizontal intake regarding Gordons' findings at prototypes.

The general outcome of the above described study of the available results of basic investigations is the possibility of establishing a simple relationship between the relative critical submergence and the induced circulation in the form of an especially defined swirl parameter (combined dimensionless number) together with the consideration of the influence of intake direction.

The developed prediction formula (Equation (4.5)) is valid for a 'moderate' circulation and for Froude numbers of the intake within a range representing normal design practice (see also Figure 4.16).

4.5. CONSIDERATION OF APPROACH FLOW GEOMETRY

The application of Equation (4.5) presupposes the knowledge of the governing circulation constant (c) as the hydraulic consequence to the approach flow geometry. The problems of getting sufficient information about this important parameter were discussed in Chapter 2. In accordance with the introductory remarks to Chapter 4, the results of the considerations in Section 4.4 may only be taken as a step forward in the direction of the necessary complete solution of the problem

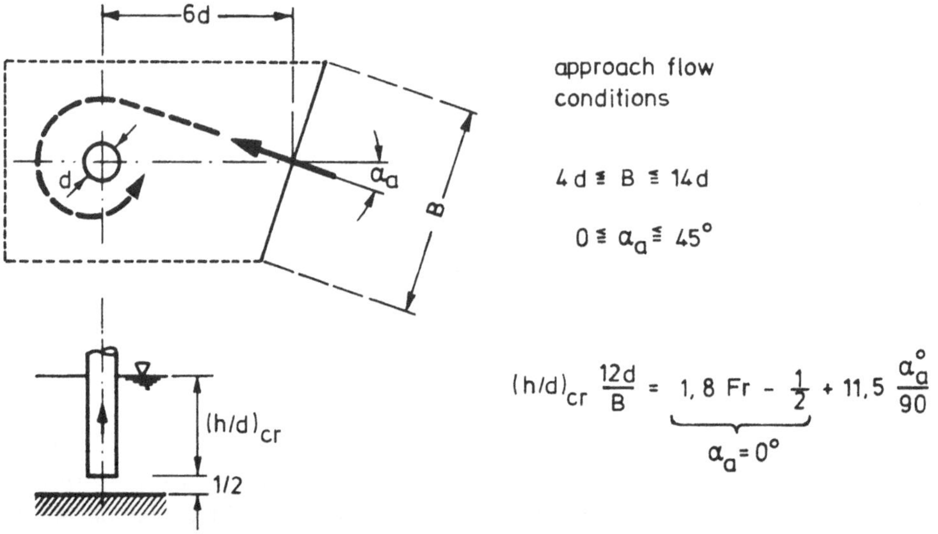

Figure 4.14. Evaluation of special tests of Amphlett (1978): approach flow angle and wall clearance versus critical submergence.

in question. But this fundamentally depends on the validity of the evaluated test results and on the possibilities of an exact calculation of the induced circulation. Nevertheless, in a lot of practical cases the application of the limited test results shown in Section 4.2 may be adequate.

Besides the h_{cr} = f(c)-relation, Amphlett (1978) gives a further evaluation of his experiments showing the dependence of the critical submergence on the approach flow angle produced with a special vane apparatus and on the wall clearance respectively the width of the approach flow area.

Figure 4.14 summarizes the generalized details of the test and includes an analytic expression which approximates the test results (see Knauss, 1983).

Further investigations of this kind are not available until now. Bypassing the problem of a direct determination of induced circulation similar research may be of some advantage for design practice.

4.6. INFLUENCE OF SPECIAL GEOMETRICAL CONDITIONS

One of the most important measures to reduce submergence requirements is a reduction of wall clearance as schematically shown in Figure 4.15.

At a certain distance from the surrounding walls different locations of the intake require an equal maximum submergence in order to avoid air entrainment. An intake impounded in the wall needs the minimum submergence. Between these limiting values the critical submergence increases linearly with increasing wall clearance.

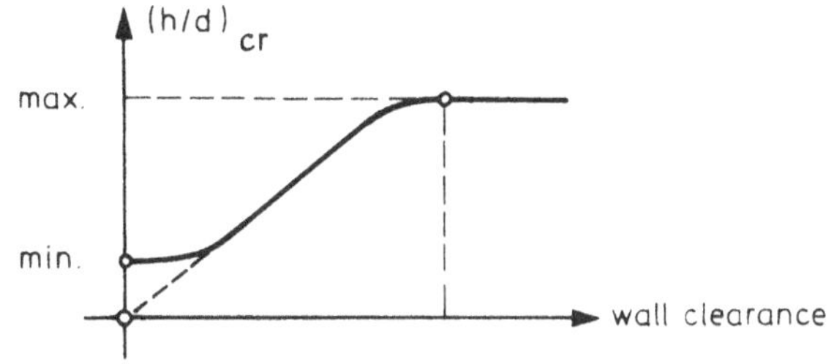

Figure 4.15. Critical submergence versus wall clearance.

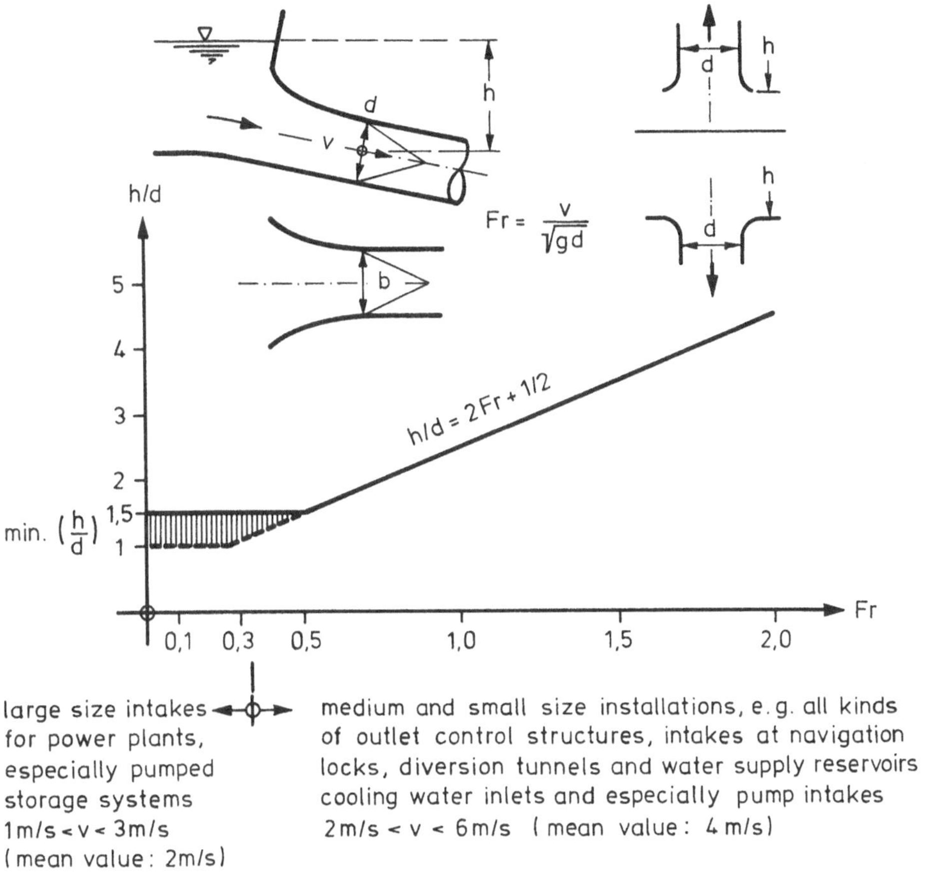

Figure 4.16. Recommended submergence for intakes with proper approach flow conditions, but without use of special devices for vortex suppression.

No information about the influence of other entrance profiles on the critical intake conditions is available. Besides the well investigated circular one, the application of the results of research to other than the circular entrance section requires the experimentally proved introduction of a characteristic dimension describing the profile under consideration similar to the use of the diameter of the circular cross-section. Due to the intallation of screens and gates at many intakes especially for power plants a prismatic entrance section is preferred. A quadratic profile or a rectangular one, which is not far away from the square shape, may be best represented by the height of the cross-section.

4.7. SUMMARY AND RECOMMENDATION

The formulation and application of an analytical prediction of the critical intake submergence as a main design tool were discussed. The final conclusion is that in most of all cases, the critical intake conditions cannot be determined with sufficient accuracy. Fortunately, submergence is not the only remedy to improve intake conditions. Providing suitable approach flow properties in order to minimize circulation may be some times much more efficient. An elimination of the sources of vortex formation has to be considered as the primary measure at intake design (see Chapter 6). The calculation of submergence requirements then may be done by an application of the recommendations given in Figure 4.16, which was developed by an evaluation of minimum design submergences at well operating prototypes as given by Gordon (1970), Pennino & Hecker (1979), Hecker (1981) and Knauss (1983). For large size intakes ($Fr \leq 1/3$) at power plants, especially at pumped storage system a submergence depth of 1 up to 1.5 times the intake height (or diameter) is recommended. For medium and small size installations ($Fr \geq 1/3$), especially at pump sumps submergence requirements may be calculated using the formula given in Figure 4.16. The recommendations are valid for intakes with proper approach flow conditions. Investigating and examining the design of an intake structure by a hydraulic model the above suggestion of minimum submergence may be taken as the aim of the final improvement of the construction. With especially well controlled approach flow conditions, with a suitable dimensioning and location of the intake relative to its surroundings and with use of proved vortex suppression devices submergence requirements may be reduced below the recommended limit (see Chapters 6.1 and 6.2).

Modelling of vortices and swirling flows 5

K.G.RANGA RAJU & R.J.GARDE
University of Roorkee, India

5.1. INTRODUCTORY REMARKS

Swirling flow at intakes – leading, quite often, to the formation of air-entraining vortices – is the result of the complex interaction between the geometry of the intake chamber and the approach channel, the flow velocity and the liquid properties such as surface tension and viscosity. The flow near an intake is quite complex and is hence not readily amenable to theoretical solution, except possibly in the case of an idealised shape of the intake chamber. Design of intakes to be free of objectionable circulation is, therefore, based either on past experience and 'rules-of-thumb' or on physical-model studies.

5.2. DIMENSIONLESS PARAMETERS

Considering an intake of the type shown in Figure 5.1, the critical submergence, h_{cr}, (defined as one which is adequate to prevent objectionable circulation in the intake or to prevent formation of air-entraining vortices) may be expressed by

$$h_{cr} = f\,(V, d, D, r, Z, \Gamma, g, \sigma, \mu, \rho) \qquad (5.1)$$

where

V = average velocity of flow through the intake of diameter d
r = radius of rounding of the bellmouth entry (if the intake has one)
D, Z = dimensions as shown in Figure 5.1
Γ (parameter) = average circulation over the depth at a distance D/2 from the centre line of the intake
g = gravitational acceleration
ρ = mass density of the liquid
σ = its surface tension
μ = its dynamic viscosity.

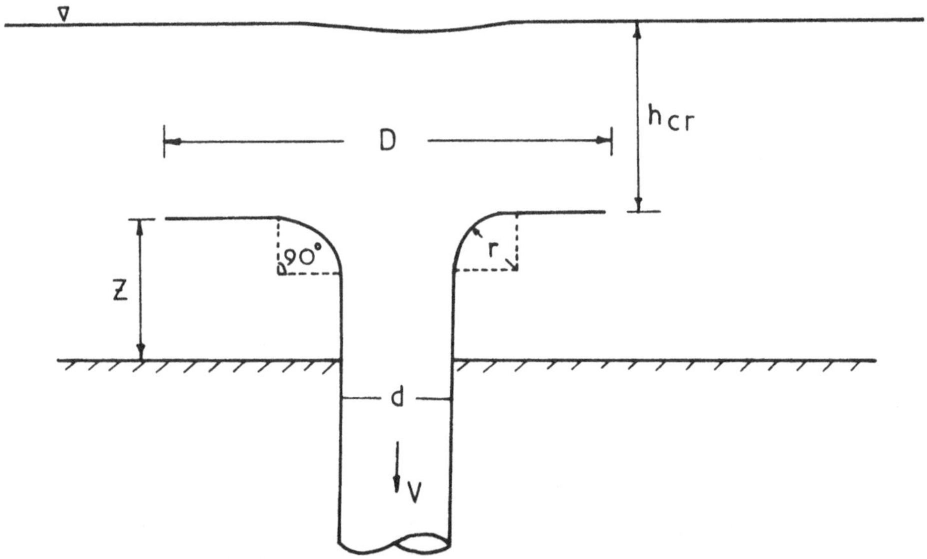

Figure 5.1. Diagrammatic sketch of a vertical intake.

Circulation, Γ, is dependent on the characteristics of the approach flow, the geometry of the intake chamber and the discharge. In the case of horizontal intakes, the reference circulation, Γ, may need to be defined along some other curve, but it is immaterial what that curve is so long as the same curve is used consistently all the time. If all the geometric parameters of the intake pipe, intake chamber and approach channel are introduced into Equation (5.1), the parameter Γ could be deleted from this equation since it is then a dependent parameter and would not figure on the right-hand side of the equation. It is convenient to use during this discussion a form of Equation (5.1) in which Γ is absent.

By way of dimensional analysis Equation (5.1) can be reduced to

$$\frac{h_{cr}}{d} = f\left(\frac{d}{D}, \frac{d}{r}, \frac{d}{Z}, \Pi_4, \Pi_5, \frac{V}{\sqrt{g\,d}}, V\sqrt{\frac{\rho d}{\sigma}}, \frac{Vd\rho}{\mu}\right) \tag{5.2}$$

where

Π^4 and Π_5 = additional dimensionless numbers (ratio of lengths) obtained as a result of introduction of additional length dimensions to eliminate Γ.

It is not important for the present discussion what the number of these additional parameters is. Equation (5.2) may be written as

$$\frac{h_{cr}}{d} = f\,(\Pi_1, \Pi_2, \Pi_3, \Pi_4, \Pi_5, \mathrm{Fr}, \mathrm{We}, \mathrm{Re}) \tag{5.3}$$

where

$\Pi_1, \Pi_2 \ldots \Pi_5$ = all ratios of two lengths and hence geometric parameters

$$Fr = \text{Froude number} = \frac{V}{\sqrt{g\,d}}$$

$$We = \text{Weber number} = V\sqrt{\frac{\rho d}{\tau}}$$

$$Re = \text{Reynolds number} = \frac{Vd\rho}{\mu}.$$

In a geometrically similar model, $\pi_1, \pi_2 \ldots \pi_5$ would be the same in the model and the prototype. Equality of h_{cr}/d in the model and prototype would, in general, require equality of Fr, Re and We between model and prototype which would be impossible to attain. The following discussion is, therefore, devoted to an analysis of the relative importance of these three parameters, which would enable the designer to decide upon the suitable modelling criteria for his study.

5.2.1. *Influence of Froude number*

Practically all studies carried out till now, with the exception of Haindl's (1959), have indicated Fr to be one of the important parameters influencing the critical submergence. This is understandable because the phenomenon is influenced by gravity and water surface is deformed in the process of vortex formation. It is therefore customary to base the model study of vortex formation on Froude number similarity though departures (or corrections to model results) are made to account for the possible distortion likely to be introduced by the non-constancy of Re and We in the model and the prototype.

5.2.2. *Influence of Weber number*

The role of the Weber number appears to be relatively less understood in spite of a few studies having been carried out recently to study its influence on vortex formation. Yildirim & Jain (1979) found that the use of a surface layer of smaller relative density than the liquid being withdrawn inhibits the formation of air-entraining vortices. Unfortunately, the values of surface tension for the liquid used in the surface layer are not available in their paper. The theoretical analyses of Hughes (1975) and Yildirim & Jain (1981) yield somewhat contradictory results. While Hughes contends that surface tension would not be important in case of a weak dimple, Yildirim & Jain conclude that surface tension would be important only at a low value of circulation (it may be noted that low circulation generally implies formation of a weak dimple at the water surface). By experimenting with liquids of different surface tensions, Jain et al. (1978) showed that vortex formation in case of vertical intakes is unaffected by Weber number so long as $We > 11.0$. It is worth noting that the above value was the lowest Weber number achieved

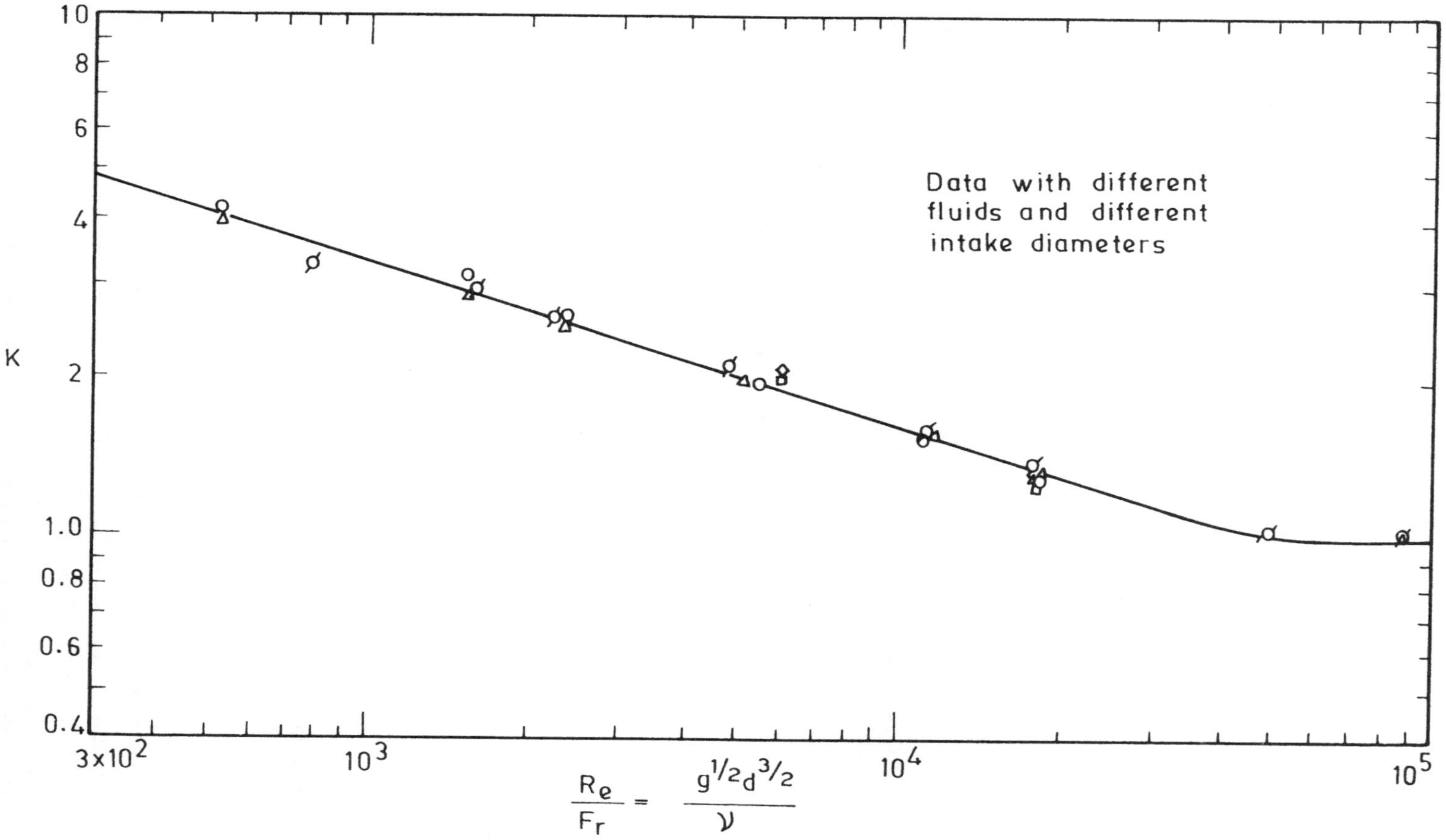

Figure 5.2. Variation of K with $\frac{Re}{Fr}$ (Jain et al., 1978).

in the experiments of Jain et al. (1978) and the limiting Weber number (below which it affects vortex formation) being lower than 11.0 cannot be ruled out. Nevertheless, in the absence of any other convincing data, it may be assumed that Weber number can be dropped from Equation (5.3) so long as its value is greater than 11.0.

5.2.3. *Influence of Reynolds number*

There are conflicting views on the role of Reynolds number in vortex formation. While Haindl (1959) believes that constancy of Reynolds number in model and prototype is essential for similitude in vortex formation, many others argue in favour of constancy of Froude number. Although Quick (1961) advocated the use of Froude number for similarity, he cautioned that the generation of swirl necessary to produce a vortex is not always independent of Reynolds number. Daggett & Keulegan's study (1974) on vortex flow through orifices showed that viscous effects become unimportant for

$$\frac{Q}{\nu d} > 2.5 \times 10^4$$

where

Q = discharge

ν = kinematic viscosity of the liquid.

and similarity is achieved by equality of Froude number. Since the velocity in orifice flow is not independent of the submergence (while they are independent of each other in pump intakes), the foregoing limit may not apply directly to pump intakes. Anwar et al. (1978) and Anwar & Amphlett (1980) contend that vortex formation in horizontal and vertical intakes is unaffected by Reynolds number when the Reynolds number

$$\frac{Q}{\nu h} > 2 \times 10^4$$

where

h = submergence of the intake.

A close scrutiny of the foregoing papers indicates that the logic leading to the above conclusion is elusive and it is felt that an analysis on lines different from that presented in these papers is necessary to establish these limits clearly.

Jain et al. (1978) used liquids of different viscosities in their study on vortex formation at vertical intakes. They found that the critical submergence in case of such intakes can be expressed by the equation

$$K\frac{h_{cr}}{d} = 5.6\ N_1^{0.42}\ Fr^{0.50} \tag{5.4}$$

where

$$N_1 = \frac{\Gamma\, h_{cr}}{Q}$$

K = function of Re/Fr as shown in Figure 5.2.

This figure shows that K is independent of this ratio for

$$Re/Fr \geq 5 \times 10^4$$

Since

$$Re/Fr = \frac{g^{\frac{1}{2}} d^{\frac{3}{2}}}{\nu}$$

it means that viscous effects are negligible if

$$\frac{g^{\frac{1}{2}} d^{\frac{3}{2}}}{\nu} \geq 5 \times 10^4$$

The tendency of some modellers to use a velocity greater than that given by the Froude number similarity criterion is a recognition of the role of Reynolds number and an attempt to account for it rather arbitrarily. It must be emphasised that the influence of Reynolds number is not fully understood at present and the differences in the various modelling techniques (discussed subsequently) stem from differences in the interpretation of the relative importance of Froude and Reynolds numbers.

5.3. EXPERIMENTAL TECHNIQUES

Since a vortex phenomenon is normally studied with the help of physical models, some discussion concerning the experimental techniques would be relevant at this stage. Visual observation of the air core or of the vortex by introduction of dye only yields qualitative results. Miniature current meters have been used to measure swirl velocities, but it is likely that, despite their small size, they disturb the flow to the extent that the stability of the vortex may be affected. Chang (1976) has reviewed some of the methods commonly used for measuring the circulation or the swirl velocities.

In the rotating prism method (see Figure 5.3) the impurities normally present in the water are illuminated by a high-intensity source S. When viewed through the telescope T, these particles travelling along streamlines would be seen crossing the field of view. If a rotating optical prism is interposed, the suspended particles viewed through the telescope will appear stationary at a certain speed of the prism. From this critical speed the velocity component of the particle perpendicular to the axis of rotation can be computed. The tangential velocity at point C is being measured in Figure 5.3; other components can easily be measured similarly by

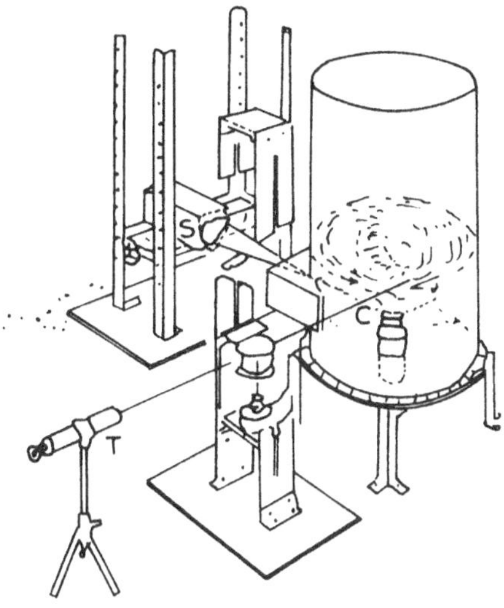

Figure 5.3. Rotating prism arrangement (Chang, 1976).

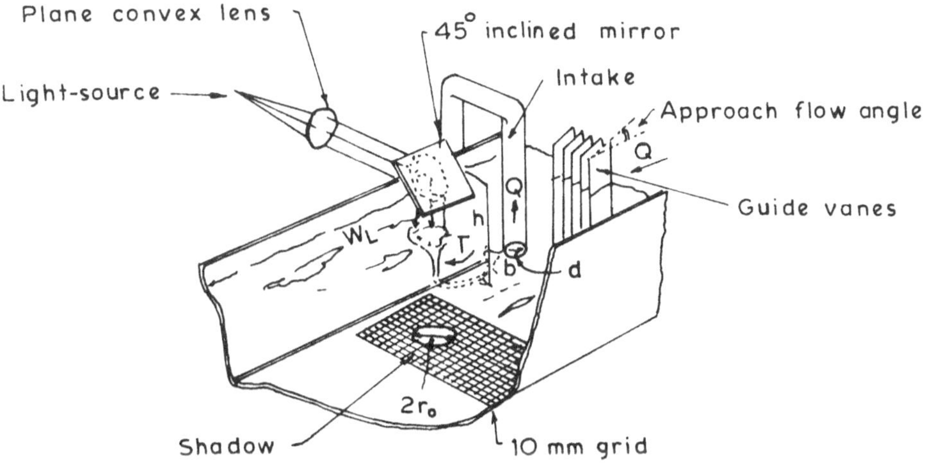

Figure 5.4. Diagrammatic set-up of refracted-ray method (Anwar & Amphlett, 1980).

rotating the prism axis. The technique has been tested up to velocities of 3.0 m/s.

In the refracted ray method the size of the shadow of the surface depression of the vortex cast on the floor by an incident beam of light is measured; see Figure 5.4. The circulation Γ is found to be related to the radius r_o of the shadow and the total water depth h_o by the equation

$$\Gamma = 0.86 \sqrt{\frac{g}{h_o}}\, r_o^2 \tag{5.5}$$

The radius r_0 is generally measured with the help of grid markings on the floor.

In the method of streak photography particles such as neutral density beads, aluminium powder or dye globules are injected into the vortex flow and the water surface photographed at regular intervals. The series of traces obtained may be used to calculate the velocity.

5.4. MODELLING CRITERIA

In the light of the discussion presented in Section 5.1, one may safely conclude that any criterion which merely advocates contancy of either Froude number or Reynolds number in the model and prototype is not satisfactory. A satisfactory criterion has to recognize that both these parameters may be important in a certain range of pertinent variables. The following are a few such criteria which merit detailed examination.

5.4.1. *Equal-velocity criterion*

Denny (1956) and Denny & Young (1957) advocate testing of a geometrically similar model at the same velocity as that in the prototype. Figure 5.5 shows that the results from such a procedure are satisfactory in the case investigated by them. The foregoing criterion was tested by them for intakes up to 0.91 m in diameter and scales larger than 1/16. Kenn & Zenker (1967) tried to interpret the equal-velocity concept as equality of the ratio of the square of Weber number to the Reynolds number, i.e. $\frac{\mu V}{\sigma}$. But it is obvious from the discussion in Section 5.1 that both Reynolds and Weber numbers are less important than the Froude number under most conditions and a modelling criterion based only on Reynolds and Weber numbers is, therefore, not justifiable theoretically. Subsequent to the publication of Denny & Young's paper, many investigators have found that adoption of the equal-velocity concept is likely to lead to a conservative design, because tendency for vortex formation is exaggerated by the adoption of a velocity so much larger than the Froudian velocity.

Chang (1979) hypothesised that the ratio of the velocities in the model and prototype required for similarity should only be a function of the geometric scale ratio and that this ratio would be different from that given by the Froude number criterion. He was, however, not successful in deriving such a relationship quantitatively on the basis of his laboratory studies.

5.4.2. *Anwar's criterion*

The results of Anwar & Amphlett (1980) and Anwar et al. (1978) may be interpreted to mean that the model scale should be so chosen that

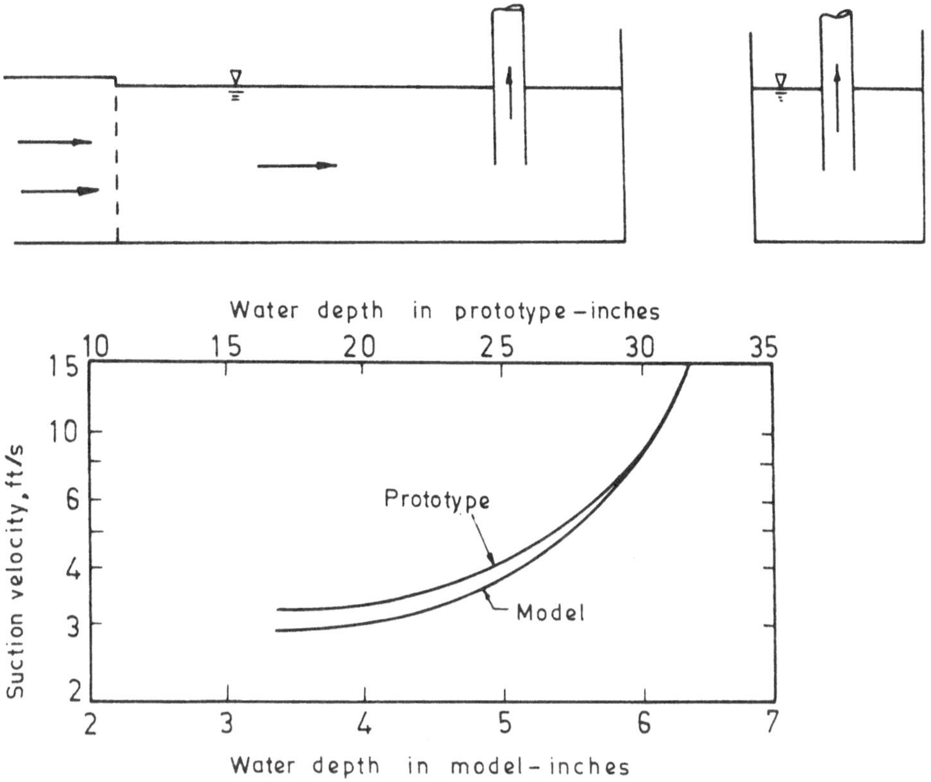

Figure 5.5. Comparison of model-prototype results (Denny, 1956) (1 ft = 0.305 m; 1 inch = 0.0254 m).

$$\frac{Q}{\nu h} > 3 \times 10^4$$

A geometrically similar model so constructed when run under the condition of equal Froude number in the model and prototype would give results directly convertible to the prototype using Froude's law of similitude.

5.4.3. *Jain et al. criterion*

Jain et al. (1978) recommend that the model be constructed as per Froude number-similarity criterion to any desired scale. Figure 5.6 enables converting the model results on critical submergence so obtained to the prototype when the Reynolds numbers in the model and the prototype are known. This correction factor is unity at high Reynolds number or, more precisely, when

$$\frac{g^{\frac{1}{2}} d^{\frac{3}{2}}}{\nu} > 5 \times 10^4$$

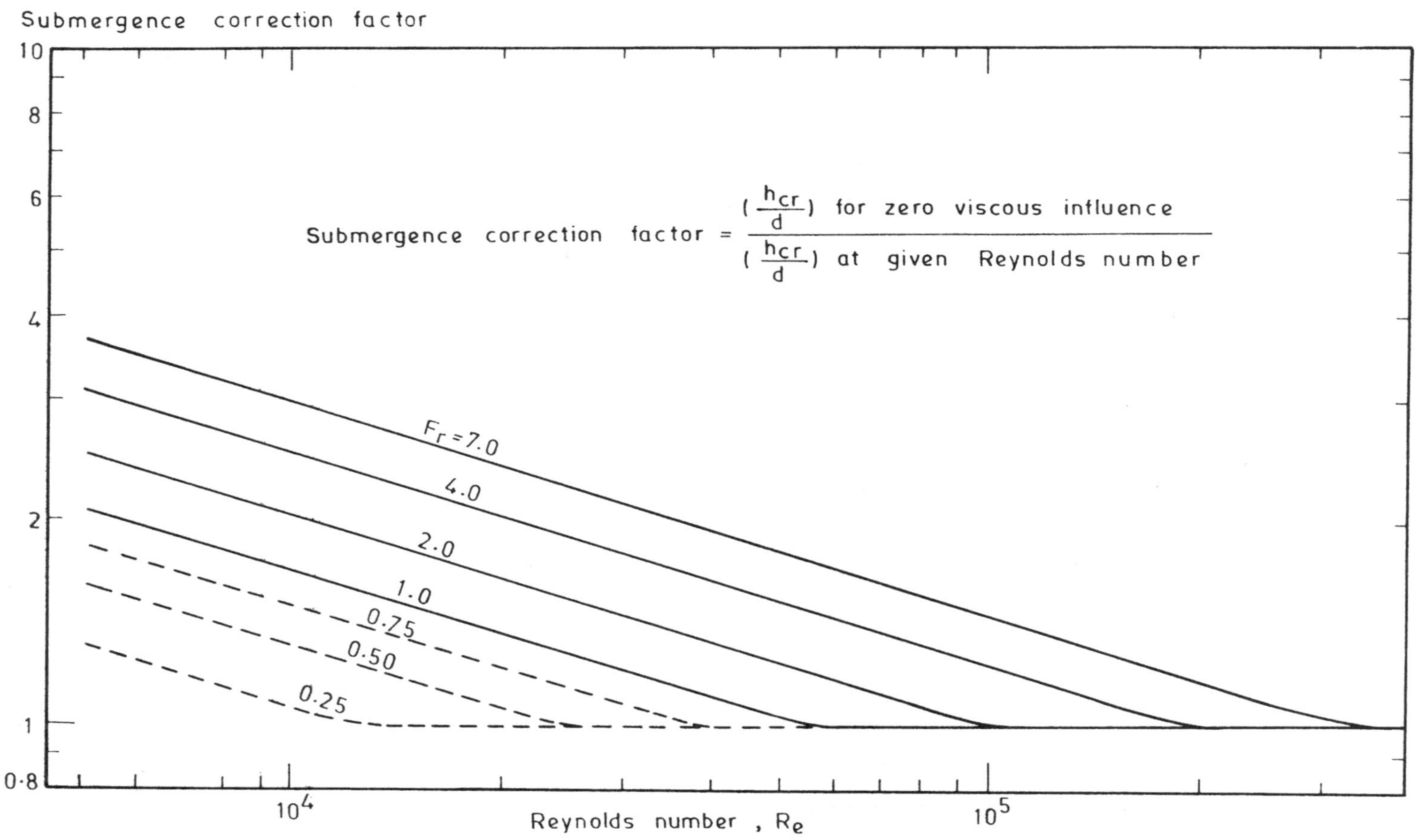

Figure 5.6. Variation of submergence correction factor with Reynolds number and Froude number (Jain et al., 1978).

It may be seen from Figure 5.2 that the constancy of K at

$$\frac{g^{\frac{1}{2}}d^{\frac{3}{2}}}{\nu} > 5 \times 10^4$$

is really based on studies with only two different diameters of intakes. It would appear necessary to check this conclusion regarding constancy of K using intakes of larger diameter also.

5.4.4. *Durgin & Hecker's approach*

Durgin & Hecker (1978) classify vortices into six different categories as listed in Table 5.1.

A typical plot of the type of vortex with $\frac{V_t}{\sqrt{g\,h}}$ for a particular case is shown in Figure 5.7. As expected, the severity of the vortex increases with increase in $\frac{V_t}{\sqrt{g\,h}}$.

Durgin & Hecker contend that the effect of Reynolds number on vortex formation drops off asymptotically and that it is therefore not possible to choose a model scale in which viscous effects are totally absent. They recommend that studies be carried out at different Froude and Reynolds numbers using water at different temperatures. If the type of vortex is observed under all these conditions, lines of approximately constant severity of vortex may be drawn. One could then extrapolate the results to the prototype judiciously. Figure 5.8 shows the technique adopted by them in making predictions in case of the cooling system of a nuclear reactor. They predicted, on the basis of Figure 5.8, that Type 4 vortex may be expected in the prototype. The technique, though sound, is cumbersome and requires additional facility for running water at different temperatures through the model.

Padmanabhan & Hecker (1984) contend that it is important to take into account

Table 5.1. Description of different vortex types (Durgin & Hecker, 1978).

Vortex type	Description
1	Coherent surface swirl
2	Surface depression
3	Coherent dye core
4	Suction of slightly buoyant particles
5	Air bubbles pulled into intake
6	Full air core to intake

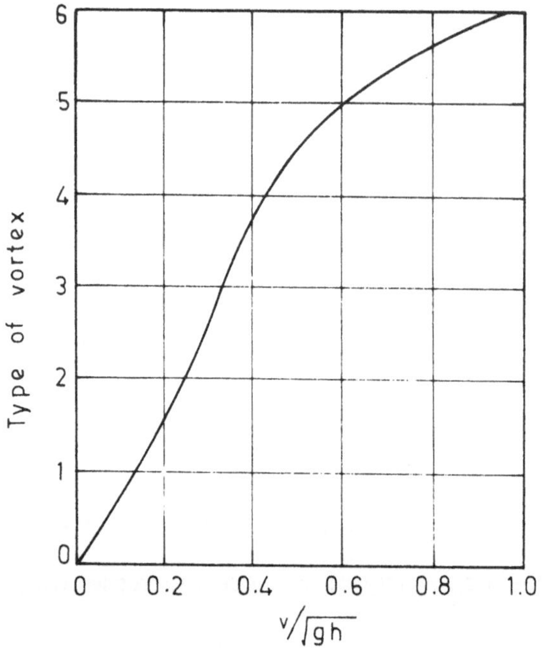

Figure 5.7. Relation between type of vortex and $v/\sqrt{gh}$ (Durgin & Hecker, 1978).

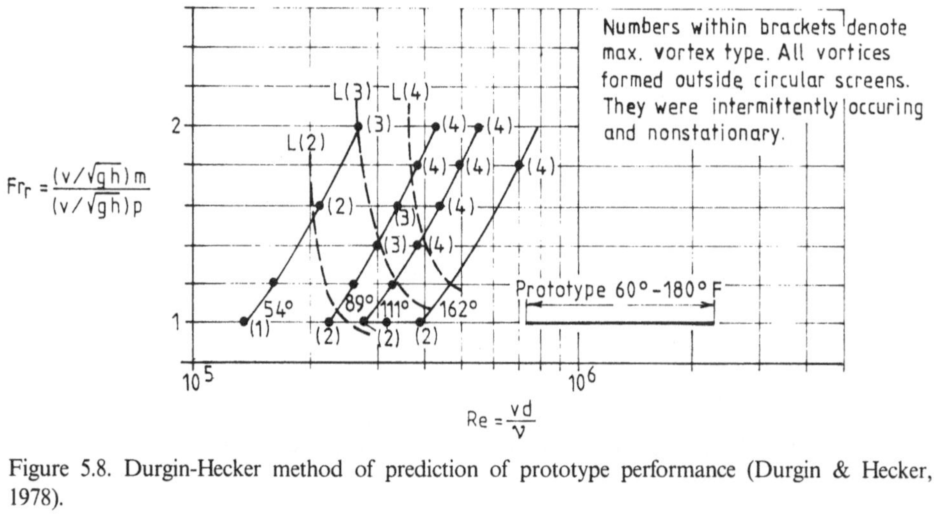

Figure 5.8. Durgin-Hecker method of prediction of prototype performance (Durgin & Hecker, 1978).

the duration for which a type of vortex occurs in classifying the strength of the vortex. The vortex strength exceeded 10% of the time may be a better parameter for classification than the strength noticed once in a while in a time-dependent phenomenon such as vortex formation. Using such a definition of vortex strength Padmanabhan & Hecker find that the vortex strength is independent of Reynolds number if the radial Reynolds number is greater than or equal to 3×10^4; see Figure

8.2. They also recommend use of a Weber number $\rho hv^2/\sigma$, in the model greater than 10^4 to obtain independence from the effects of surface tension.

5.5. VERIFICATION OF MODELLING CRITERIA

Not many detailed studies have been carried out with the intention of verifying the correctness of the foregoing modelling criteria. A notable exception is, however, the highly painstaking work carried out by Dhillon (1980) at Amritsar, India. He built models of prototypes with vertical and horizontal intakes to different scales as per the recommendations of Denny & Young (1957), Anwar et al. (1978) and Jain et al. (1978). The behaviour in the prototypes at different water levels was observed and the behaviour in the various models under these very conditions was studied. It was found that the equal-velocity approach generally indicated vortex formation in the model under conditions when no vortices were observed in the prototype. The model built as per Anwar's criterion with the further restriction that the scale should not be less than 1:20 yielded results in good conformity with prototype observations. It was also seen that the model built with an intake diameter satisfying the criterion of Jain et al., viz. that

$$\frac{g^{\frac{1}{2}}d^{\frac{3}{2}}}{\nu} > 5 \times 10^4$$

indicated no vortices under the conditions leading to vortex formation in the prototype. It may be mentioned that recent experiments on horizontal intakes (Bhargava, 1982) have shown the influence of viscosity on persist even when

$$\frac{g^{\frac{1}{2}}d^{\frac{3}{2}}}{\nu} > 5 \times 10^4$$

Hecker (1981) collected data from different parts of the world on model-prototype comparisons in so far as vortex formation is concerned. In general, it was found that modelling as per Froude number-similarity criterion gave good results. No limit on the minimum Reynolds number to be used for making model results free from viscous effects could be prescribed as a result of this study.

5.6. CONCLUDING REMARKS

The available theoretical and experimental evidence suggests that equality of Weber number in the model and prototype may be unnecessary in most situations, because the scales customarily used give a Weber number high enough to preclude surface tension effects on vortex formation. Modelling on the basis of equal velocity in model and prototype appears theoretically unjustifiable and unacceptable in the light of available data. Physical models may be built as per the Froude number similarity criterion and tested at radial Reynolds number greater than 3 10^4 to eliminate viscous influence as per Anwar et al. Despite Dhillon's finding that this criterion is satisfactory, it appears necessary to check it carefully because the basis for prescribing this limit is not very clear. The suggestion of Jain et al. that models with intake diameter such that

$$\frac{g^{\frac{1}{2}}d^{\frac{3}{2}}}{\nu} > 5 \times 10^4$$

are free from viscous effects needs to be checked in the light of Dhillon's work. The technique proposed by Durgin & Hecker appears satisfactory but is very cumbersome.

Design recommendations 6

6.1 *Intake structures*

P.RUTSCHMANN, P.VOLKART & D.VISCHER
Federal Institute of Technology, Zurich, Switzerland

6.1.1. INTRODUCTION

Vortices at intake structures produce various problems during operation of the intake. Apart from discharge losses, severe damage may occur to the construction and associated mechanical parts. Therefore, every design must prohibit swirling flow problems. There are several causes for vortex formation. These include eccentric orientation of the intake, asymmetric approach flow conditions, obstructions in the flow and non-uniform velocity distribution. Measures to prevent rotational flow must either reduce or eliminate these causes. As intake designs and approaching flow conditions vary from project to project, it is usually not possible to investigate the problem analytically. The use of a hydraulic model often proves to be advantageous in finding or improving an intake design.

6.1.2. TYPES OF INTAKES

Intakes are normally large size installations associated with power plants, pumped storage schemes, water supply systems, lock filling installations or various other dam hydraulic structures. They can be classified according to two main criteria (see also Section 1.3): the intake direction and whether the intake is constructed flush with the boundaries or protrudes into the reservoir. This classification also considers the different flows into the intake that have associated vortex formation in order to predict critical submergence for a given intake design (see Chapter 4). As intake designs always depend on their hydraulic purpose as well as on topography and other factors, myriads of different intake designs have been constructed up to the present day. It seems therefore useless and impossible to introduce a more precise intake classification.

6.1.3. IMPORTANT PARAMETERS

The equation of motion for vortex flows written in dimensionless form (see also

Section 2.1.5) allows determination of the basic parameters governing vortex occurrence. The following relevant dimensionless parameters can be used:

$$Fr = \frac{v}{\sqrt{(g\,h)}} \quad \text{(submergence Froude number)}$$

$$N_\Gamma = \frac{\Gamma\,d}{Q} \quad \text{(circulation number)}$$

$$Re = \frac{Q}{\nu h} \quad \text{(radial Reynolds number)}$$

$$h/d \quad \text{(relative intake submergence)}$$

$$We^2 = \frac{\rho\,v^2\,d}{\sigma} \quad \text{(Weber number)}$$

where

$$v = \frac{Q}{\pi}\,\frac{d^2}{4}$$

h = intake submergence
d = intake diameter
g = acceleration due to gravity
Q = intake discharge
ν = kinematic viscosity
ρ = fluid mass density
σ = surface tension force.

Any measures taken to avid vortex occurrence at intake structures must influence these parameters. For prototype intake designs the submergence depth, the intake geometry or the intake discharge (both of which affect the intake velocity) and the reduction of the circulation parameter can be influenced positively.

6.1.4. DESIGN GUIDELINES AND PREFERRED DESIGNS

The design of a vortex free intake has to consider the following points: submergence depth, intake geometry and approach flow conditions. By giving enough attention to these factors a lot of trouble and costs can be avoided in the state of preliminary design. Advice concerning prediction or reduction of the critical submergence depth are given in Chapter 4 and will not be treated here any further. The intake geometry affects the vorticity by its cross-sectional area (determining velocity) and by its streamlined shape (elimination of vortex singularity). The position relative to the topography and the topography itself may prohibit asymmetric inflow conditions which can be the origin of a vortex formation. If these points can not be optimized in the design state for various reasons additional, secondary measures e.g. special antivortex devices can effect the elimination or suppression of surface vortices. The following vortex eliminating measures are possible:

Elongation of streamlines between the intake and the free water surface. This can be achieved with:

— higher minimum water level, i.e. higher submergence,
— lower level of intake structure,
— changing of the inflow direction,
— horizontal, protruding roof at the top of the intake opening.

Elimination of approach flow non-uniformities. This can be achieved with:

— uniform velocity distribution by means of appropriate elements,
— additional elements to direct flow toward intake,
— elimination of secondary flow and asymmetrical boundary conditions e.g. by injectors,
— streamlining of piers, etc.,
— varied cross-sectional area at intake,
— partial closure of gates,
— slightly accelerated approach flow.

Special vortex suppression devices. Examples are:

— vertical walls or vortex suppressing horizontal beams,
— floating rafts in regions of strong vorticity,
— steeper inclination of the intake surroundings.

The examples of Section 6.1.6 show designs that have been model tested and applied in the prototype situation. They represent only a representative sample of the incredibly large number of intake designs. The supplementary comments enhance understanding of the operation of the intake.

As vortex phenomena are rather complex it is not always possible to exclude prototype vortex appearance for a given design. In these cases, and for the investigation of the functionality of special anti-vortex devices, hydraulic model tests still enable the best prediction of prototype vortex occurrence to be made.

6.1.5. RECOMMENDED DESIGN CORRECTIONS

Design corrections to an already operating intake are always difficult to achieve. For this reason hydraulic model investigations are an appropriate way to find a satisfactory solution. In the duration of model testing the whole range of vortex preventing or suppressing measures can be applied (see Section 6.1.4). Application of these measures to operating prototype intakes may be undesired or even impossible. However, some prototype improvements are possible, but will probably cause operational restrictions. These possibilities are:

— raising the minimum water level,

— decreasing of the intake discharge with partial closure of gates for critical conditions,

— modification of the surrounding terrain,

— installation of floating rafts or balls in the vortex core region to diminish rotational flow or to prevent air entrainment.

The next chapter presents some examples of anti-vortex devices designed in the preconstruction period.

6.1.6. EXAMPLES

The following figures give a sample of anti-vortex devices presented in engineering literature. The numbers in the square brackets refer to the list of references. The sequence of the examples goes from circulation dominated vortices such as bath tub vortices with vertical inflow direction to flow velocity dominated vortices such as those experienced at power-house intakes with horizontal inflow direction. The symbols used in the examples refer to the following figure.

6.1.7. CONCLUDING REMARKS

As is known, the appearance of intake vortices is a very sensitive phenomenon and usually depends on more than one hydraulic parameter. Model tests can be valuable in terms of assisting classification of the approaching flow, and for finding the most successful measures for preventing or suppressing vortices. Moreover, the frequency (of appearance) of critical boundary conditions (e.g. low water levels), the intake type, and therefore the necessary degree of vortex elimination (e.g. total vortex elimination or simply elimination of air core vortices), plays a major part in indicating the appropriate preventative measures. In some cases minor modifications to achieve a uniform approaching flow velocity distribution such as topography or dam slope corrections near the intake may suffice to avoid operational vortex problems. This may avoid the necessity of constructing expensive vortex suppressors which seldom come into operation.

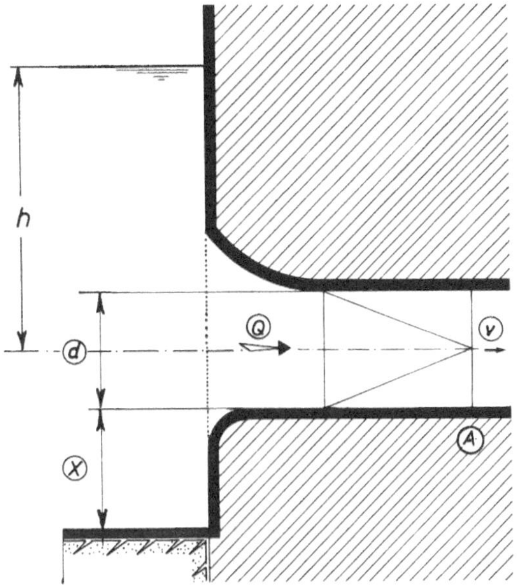

Definition diagram for symbols:

A = minimum cross-sectional area at the intake

d = minimum diameter at intake with circular cross-section

h_{max} = maximum submergence depth

h_{min} = minimum submergence depth

Q = intake discharge

v = intake velocity in the minimum cross-section

x = distance of intake above basin bottom.

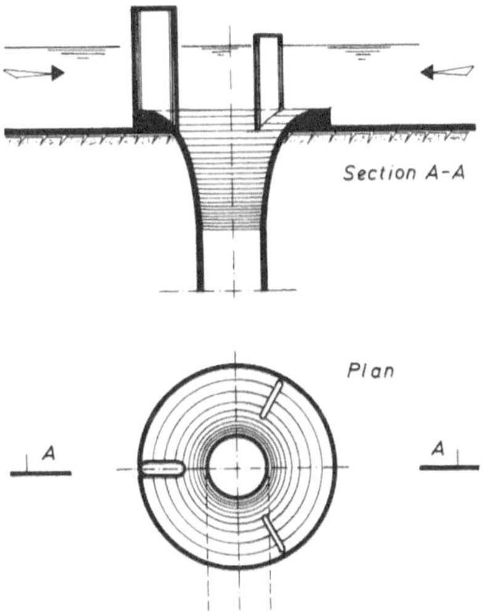

1. Del Valle Dam (USA): Bath tub vortex elimination with radial piers. Aeration of the spillway elbow is achieved with an additional air duct connecting the aeration pier and the deflector (Bucher & Amorocko, 1968).

h_{max} = 5.95 m

h_{min} = 0 m

d = 8.53 m

Q = 1250 m^3/s

v = 21.8 m/s

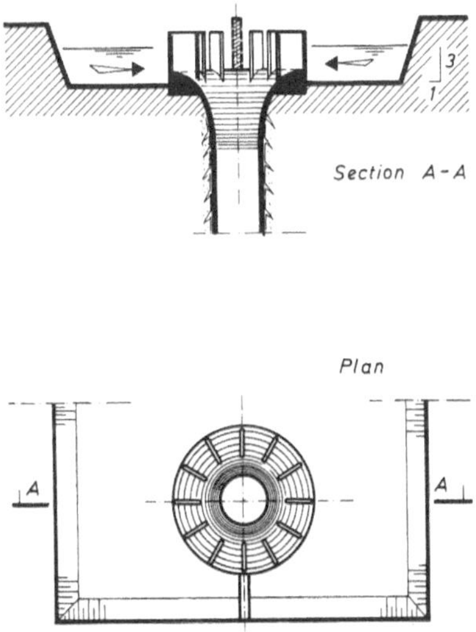

2. Radial piers supported by topography modifications and a vertical vane. Construction only investigated in the model (Allen & Taylor, 1959).

h_{max} = 1.52 m

d = 3.05 m

Q = 90.6 m^3/s

v = 28.8 m/s

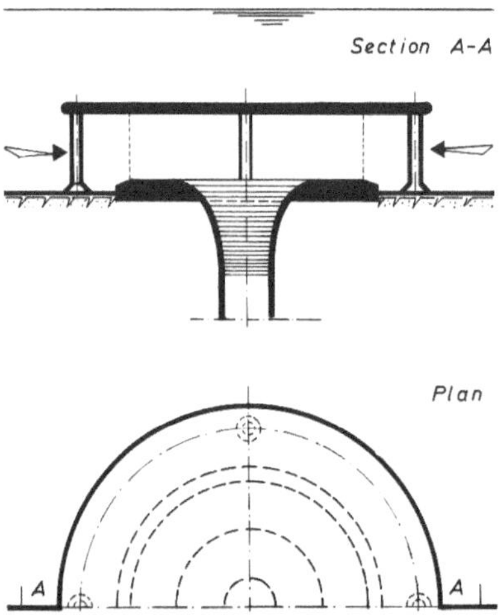

3. Principal proposal to reduce bath tub vortices in pump sumps by using a circular cover plate. The similarity to a morning glory spillway is evident (Prosser, 1977).

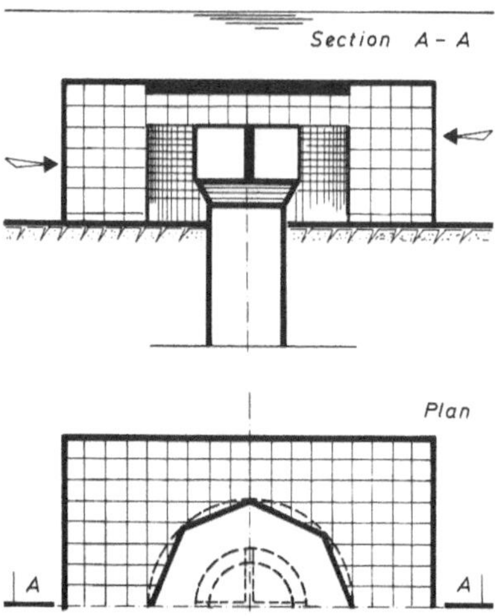

4. J.M.Farley (USA): Combination of measures for suppressing vortices (cover plate, guide vanes and grids). The measures were tested in a model with 50% increased velocities (Elder et al., 1978).
h_{min} = 1.48 m   Q = 0.272 m^3/s
h_{max} = 1.96 m   v = 3.66 m/s
d = 0.36 m

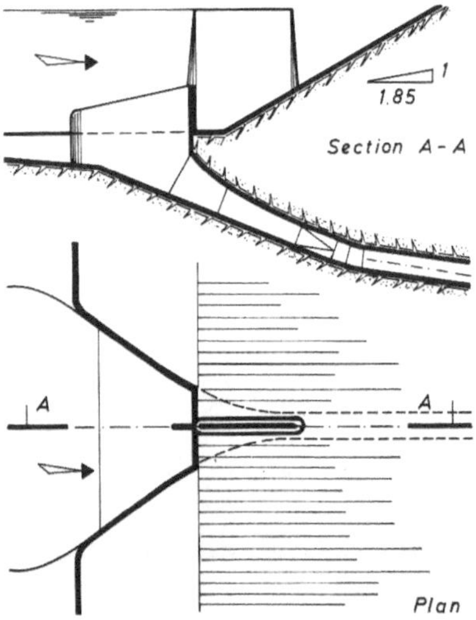

5. Mosul (Iraq): Construction for generation and pumping cycle. Rotation prohibited by vertical vanes (VAW Reports).
h_{min} = 5.16 m   Q = 80.3 m^3/s
h_{max} = 26.36 m   v = 6.4 m/s
d = 4.00 m

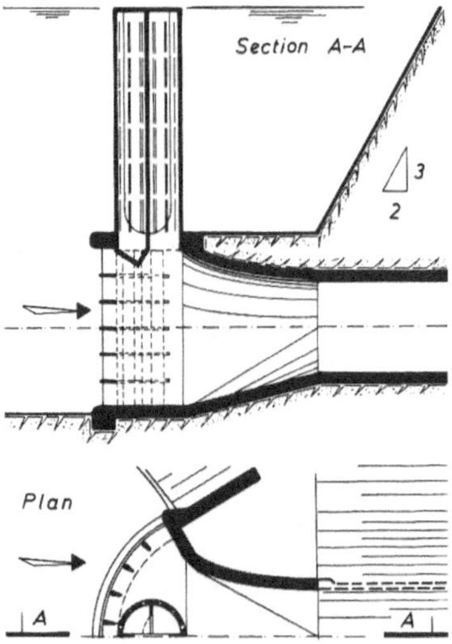

6. El Cajon (Honduras): The tower stabilizes vortex formation and reduces rotation by sucking in water from the dead water zone. The structure was model tested with increased velocities (Schulthess, 1985; VAW Reports).
h_{min} = 9.00 m   Q = 1780 m^3/s
h_{max} = 41.50 m   v = 15.7 m/s
d = 12.00 m

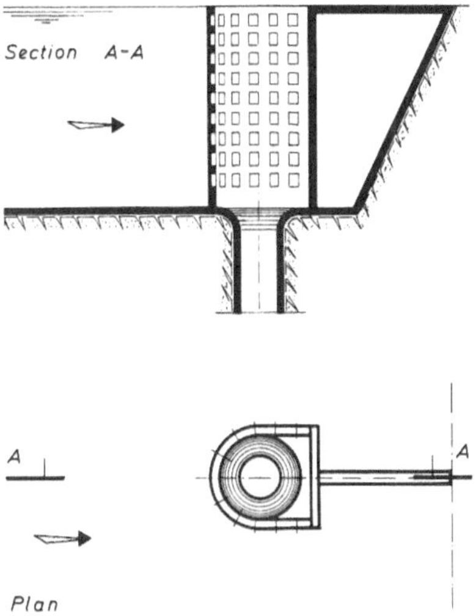

7. Rakoon Mountain (USA): Vortex suppression achieved with a vertical wall and the elimination of dead water zones. Model tests indicated high sensitivity to changes in topography (TVA Report).
h_{min} = 22.9 m Q = 673 m^3/s
h_{max} = 66.1 m v = 7.1 m/s
d = 10.67 m

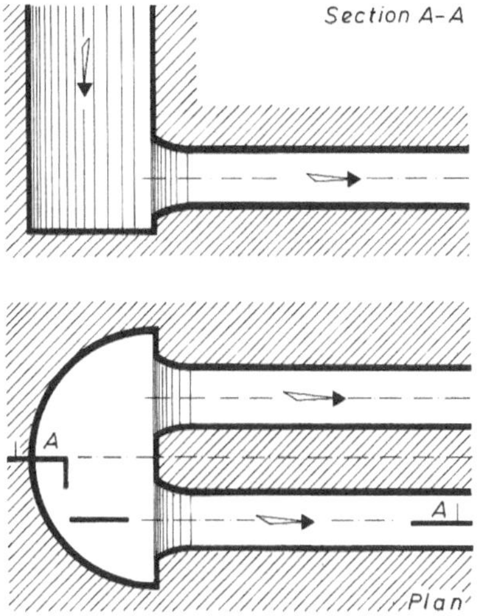

8. Cristal Dam (USA): Vortex elimination achieved by the elimination of dead water zones at the bottom of the tower (Burgi & Fujimoto, 1973).
h_{min} = 54.67 m Q = 22.7 m^3/s
h_{max} = 71.43 m v = 7.7 m/s
d = 1.37 m

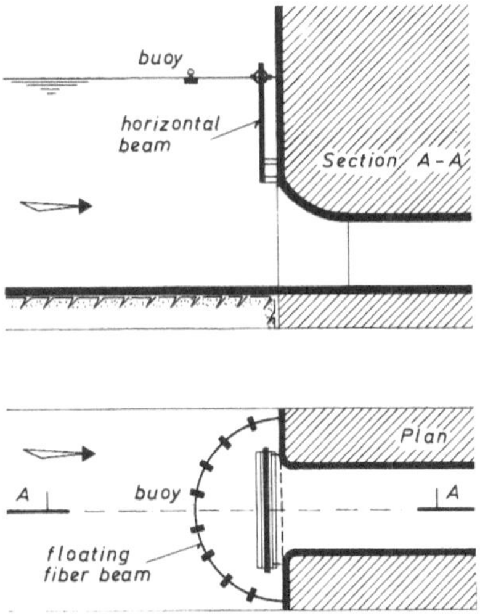

9. Gevelinghausen (Germany): Vortex rotation was eliminated with the help of buoys and a horizontal beam (Bretschneider, 1981).
h_{max} = 4.82 m
A = 4.5 × 11.4 m
Q = 92 m^3/s

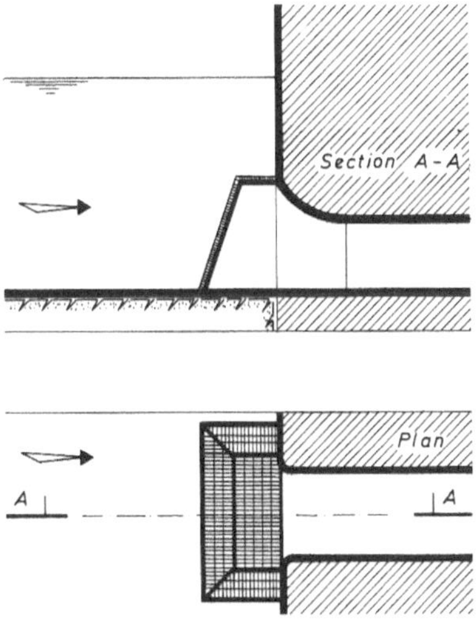

10. St. Anthony Falls (USA): Improvement of the inlet by introducing trash racks was necessary because of a fatal accident (Ables, 1979).
h_{max} = 7.99 m
A = 2.44 × 3.05 m

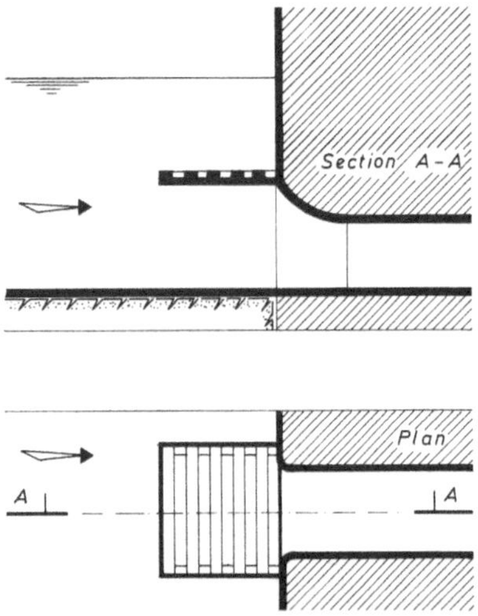

11. Principal proposal to reduce vortices in pump sumps with the help of a vortex suppressing grid. The similarity to power intakes is evident (Prosser, 1977).

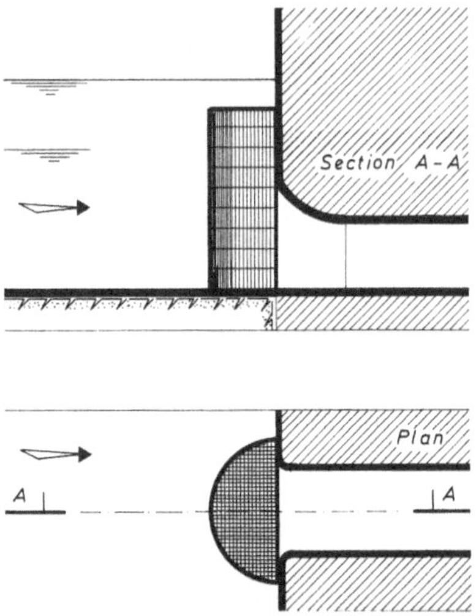

12. Grand Coulee (USA): Vortex reduction achieved due to additional losses and the elimination of dead water zones. Model tests with increased velocity (up to three times) produced stronger vortices in the model than occur in the prototype (Dexter & Zeigler, 1978).

h_{min} = 20.00 m Q = 878.00 m^3/s
h_{max} = 45.00 m v = 7.52 m/s
d = 12.19 m

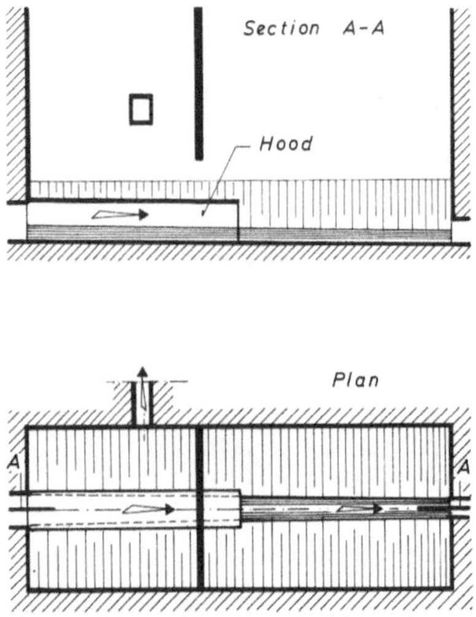

13. General design for stilling pond overflows. Only model results are avilable (Reddy & Pickford, 1974).

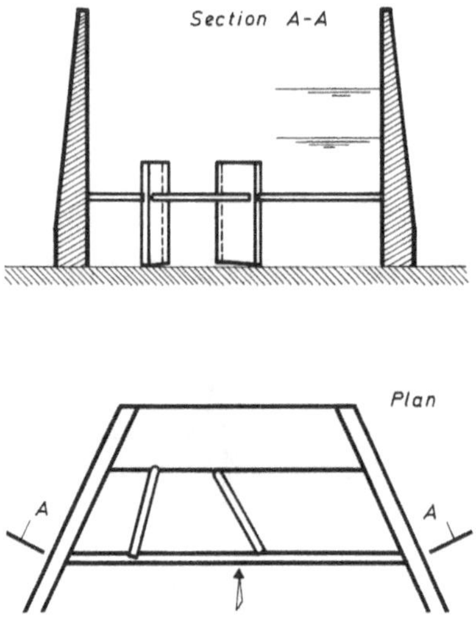

14. Mayfield Dam (USA): Vortex prevention is achieved with the help of guide vanes which produce additional turbulence (Herstrom et al., 1982).

d = 5.40 m v = 4.3 m/s
Q = 391.0 m^3/s

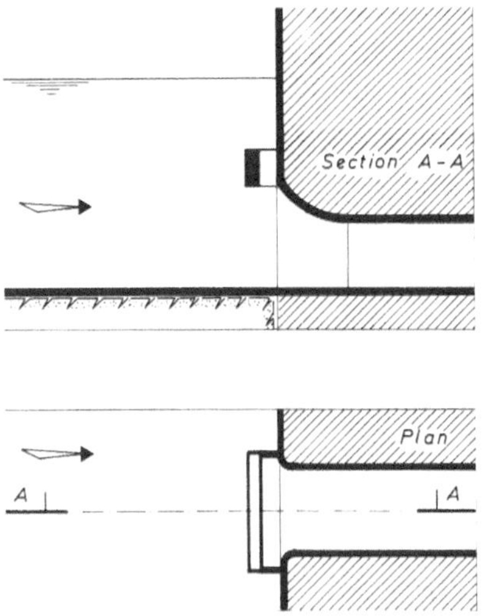

15. Kariba Dam: The baffles were positioned through the centre of the vortex. It was anticipated that rafts would be necessary if air entrainment was observed in the prototype (Linford, 1965).
h_{min} = 6.10 m   Q = 136.0 m^3/s
h_{max} = 64.00 m   v = 4.6 m/s
d = 6.09 m

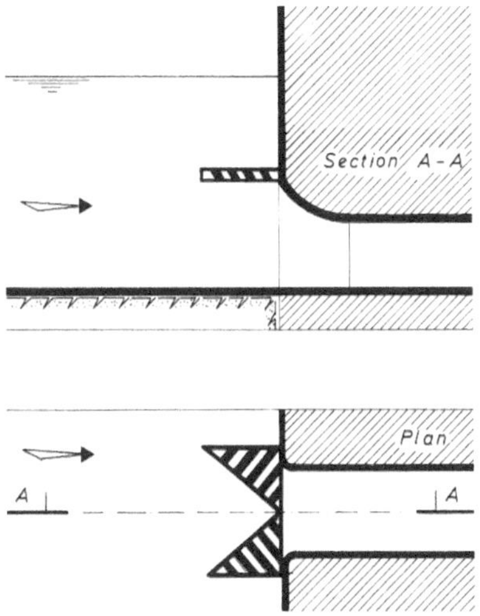

16. Dinorwich (UK): Horizontal rack elements are used to eliminate the appearance of vortices (Papworth, 1975).
h_{min} = 4.75 m   Q = 130 m^3/s
h_{max} = 18.25 m   v = 2.28 m/s
A = 57.1 m^2

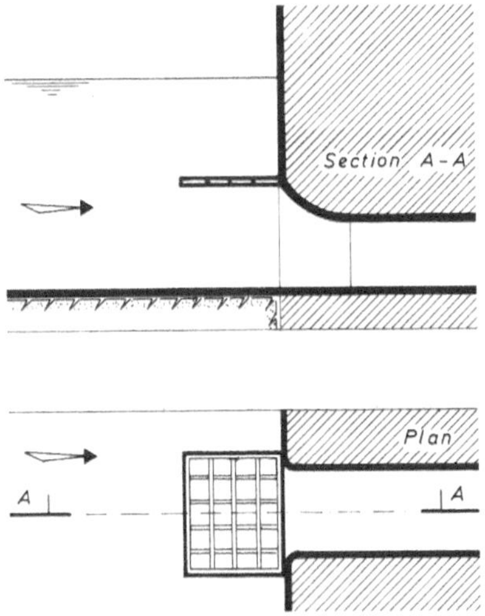

17. Mt. Elbert (USA): Rotation is hardly influenced but the rack effectively prohibits air entrainment. A raft was proposed as a second possibility for vortex suppression (Johnson, 1972).
h_{min} = 10.26 m   Q = 101.9 m^3/s
h_{max} = 19.86 m   v = 6.22 m/s
d = 4.57 m

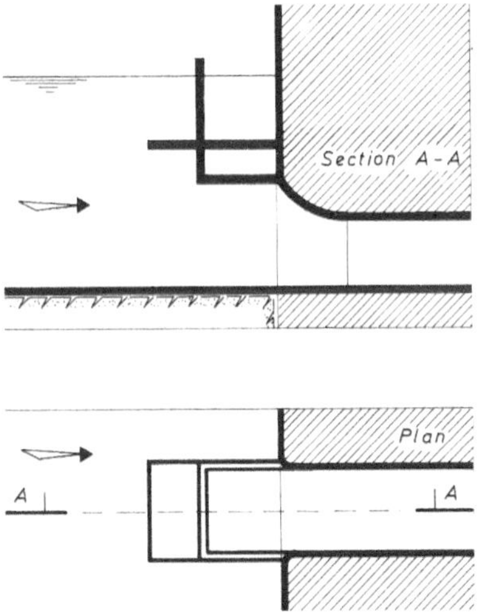

18. Principal proposals for intake designs. Vortex prevention is achieved by elongation of the flow path (Anwar, 1968b).

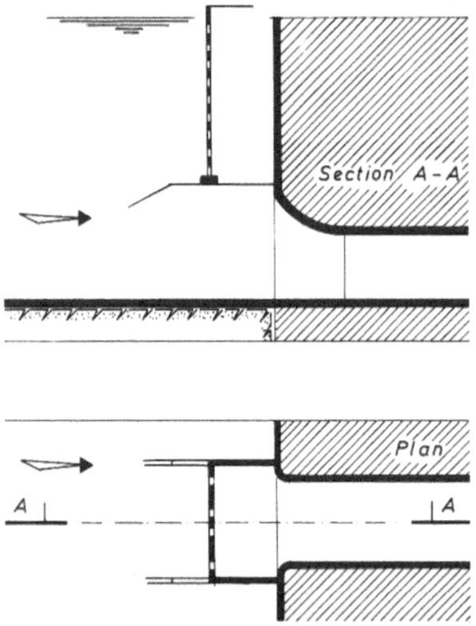

19. Luddington (USA): Vortex elimination is achieved with elongation of the streamlines and elimination of the dead water zones. Each inlet baffle wall contains 36 holes (Comninellis, 1973).

h_{min} = 12.22 m Q = 101.9 m^3/s
h_{max} = 32.64 m v = 6.22 m /s
d = 4.57 m

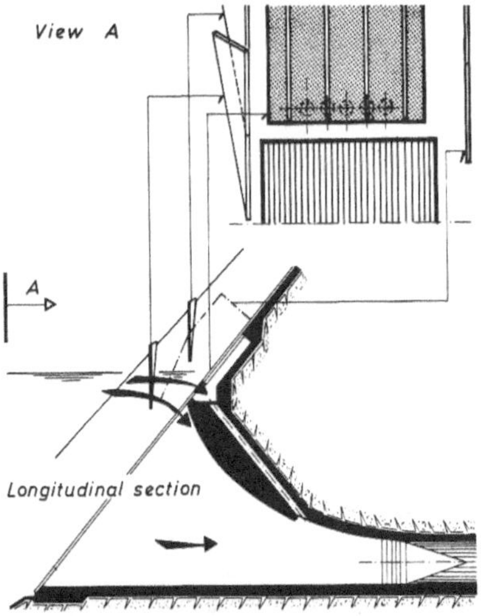

20. El Cajon (Honduras): A combination of guide walls and injector prohibits vortex formation. The injector favours a parallel streamline pattern by sucking water from the dead water region (Schulthess, 1984; VAW Reports).

h_{min} = 18.40 m Q = 120 m^3/s
h_{max} = 82.40 m v = 70 m /s
d = 4.70 m

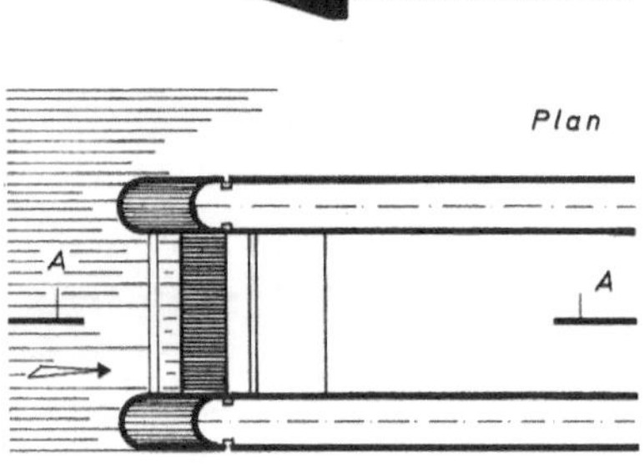

21. Bremgarten (Switzerland): The injector favours parallel flow to the inlet and prohibits flow rotation and dead water zones. The injector device served as a vortex suppressor and forced floating debris to the trash rack (Bisaz et al., 1976; VAW Reports).

h = 15.50 m Q = 100 m^3/s
A = 8.59 m^2 v = 11.7 m/s

6.2 *Pump sumps*

M.PADMANABHAN
Alden Research Laboratory, Holden, Massachusetts, USA

6.2.1. INTRODUCTION

Free-surface and subsurface vortices occur in a pump sump due to the vorticity generated within the sump area by one or more vorticity sources. Vorticity sources may be generally categorized into three fundamental types, as discussed by Durgin & Hecker (1978); namely, (a) non-uniform approach flow to the sump due to the geometric orientation of sump or approach channel or due to streaming flow patterns generated by obstructions such as intake piers; (b) existence of shear layers of high velocity gradients, including separated boundary layers which are inherently rotational; and (c) rotational wakes generated by objects or obstructions in the way of the approach flow to the sump. Item (a) is a major source of vorticity in surface vortices while item (b) is a major source of vorticity in subsurface vortices. A good design of the pump sump essentially involves reducing the contribution from these vorticity sources as far as practicable and thereby reducing the strength of the vortices generated by them to acceptable limits. An analytical determination of the strength of these vorticity sources and their combined effect on vortexing is impossible due to the complexity of the system, and hydraulic model studies are often used in verifying or arriving at an acceptable sump design.

6.2.2. CONCEPTS OF SURFACE AND SUBSURFACE VORTEX CONTROL

Controlling surface vortices is possible mainly by three methods: (a) increasing the submergence of the pump entrance from the water surface to values above critical limits that cause objectionable vortexing; (b) eliminating the approach flow non-uniformities, thereby reducing the contribution from this vorticity source to surface vortexing; and (c) providing vortex suppression devices. As regards to the first method, the submergence can be increased by lowering the sump floor and/ or pump 'bell' suction entrance or by increasing the minimum water level at which the pump is operated. Both of these options may be restrictive due to practical considerations. As regards the second method, this could be attempted by a combination of several means: namely, modification of sump geometry and location with respect to approach channel and/or use of flow distributors or velocity modification devices (head loss devices) such as screens, gratings, and perforated walls. In the third method, the rotational energy contributing to vortices is dissipated significantly with suppression devices such as horizontal gratings, cages, and rafts. This method is especially suitable if both approaches (a) and (b) above are restricted by practical considerations.

Controlling of subsurface vortices mainly involves elimination of large velocity gradients in the flow and avoiding flow separation inside the sump. To achieve this, the sump geometry should avoid sudden flow transitions and should provide sufficient wall and floor clearances from the bell entrance to avoid flow separations with large velocity gradients. Artificial devices such as floor cones, splitters, and fillets, are installed to prevent flow separation and to guide the flow into the suction bell. A good arrangement generally is derived by trial using a hydraulic model so that the intensity of induced swirl in the suction pipe would be negligible.

6.2.3. HYDRAULIC PERFORMANCE AND ACCEPTANCE CRITERIA

Before a preliminary design for a pump sump is selected, it is important to understand the hydraulic problems which affect the pump performance. The major problems of concern are:

a) Air-drawing surface vortices could reduce the pump capacity and increase the required Net Positive Suction Head (NPSH) depending upon the amount of air drawn. For centrifugal pumps, small quantities of air such as, 3 to 6% by volume at the pump location, decreases the efficiency of the pump, and when the volumetric concentration rises to 7 to 20% a substantial performance degradation would result (Murakami et al., 1968; Patel & Runstadler, 1978). Based on tests on emergency recirculation pump sumps for nuclear power stations, Padmanabhan (1982a) has shown in general that air core vortices, draw less than two percent air by volume under relatively uniform approach flow conditions within normal ranges of flows and submergences. Corresponding to submergence Froude number

$$Fr = \frac{v_D}{\sqrt{g\,h}} < 0.8$$

where

v_D = average velocity in suction pipe
h = submergence of the pipe from water surface
g = acceleration due to gravity.

However, under extremely non-uniform approach flow conditions and at higher Froude numbers, the same sump could have strong air-core vortices drawing large quantities of air, as much as 18% by volume. With the extent of approach flow non-uniformity unknown in most cases, avoiding air drawing vortices is an important criteria for sump design. It may be noted that strong subsurface vortices also could allow release of dissolved air, however, this is usually a negligible amount of air.

b) Non-air-drawing vortices, with a strong coherent core visible with dye injection or strong enough to pull debris, are also considered objectionable since they may cause vibrations due to impellers intercepting nonsymmetrical pressure fields.

c) Surface and subsurface vortices are known to contribute to swirl in the flow into the suction bell. Swirl present at the entrance to the suction bell will be transmitted to the impeller and this could induce impeller vibration and cavitation, and also could lower the pump efficiency. How much swirl is allowable? This depends on the type of pump and no definite answers are available. As a general rule, swirl angles above five degrees should be considered unacceptable (Sweeney et al., 1982).

d) Uneven flow distribution to the impeller would give rise to unbalanced loading on the impeller. Uniform velocities at the impeller location in magnitude within ±5% of the mean velocity may be considered as a requirement (Tullis, 1979).

6.2.4. TYPES OF PUMP SUMPS

A pump sump may have a vertical upward, vertical downward, or horizontal suction outlet, depending upon the type of pump. Sumps with vertical upward flow outlets are usually associated with axial flow pumps widely used in practice and can be a wet well or a dry well type as illustrated in Figure 6.2.1. The wet well type is more common, being simple and economical even though maintenance or repairs to the impeller would require dewatering the sump or removing the pump as a whole. Horizontal or vertical downward suction pipes may be long with the pump impeller located away from the sump. When longer suction pipes are involved, certain problems associated with the sump design, such as swirl generated by free-surface and sub-surface vortexing, become less important as swirl is known to decay with distance (Padmanabhan & Janik, 1980). However, additional care as regards to available net positive suction head would be necessary to account for any increased losses due to swirling flow.

A pump sump can be a single or a multiple depending on whether one or more pumps draw water from the same sump. Multiple pump sumps may have different bays or cells for each pump with a common approach channel.

6.2.5. AVAILABLE DESIGN GUIDELINES

A preliminary design of a pump sump could be obtained using the guidelines on sump dimensions and related pump position given in the Hydraulic Institute Standards (1975) or in the British Hydromechanics Research Association (BHRA)

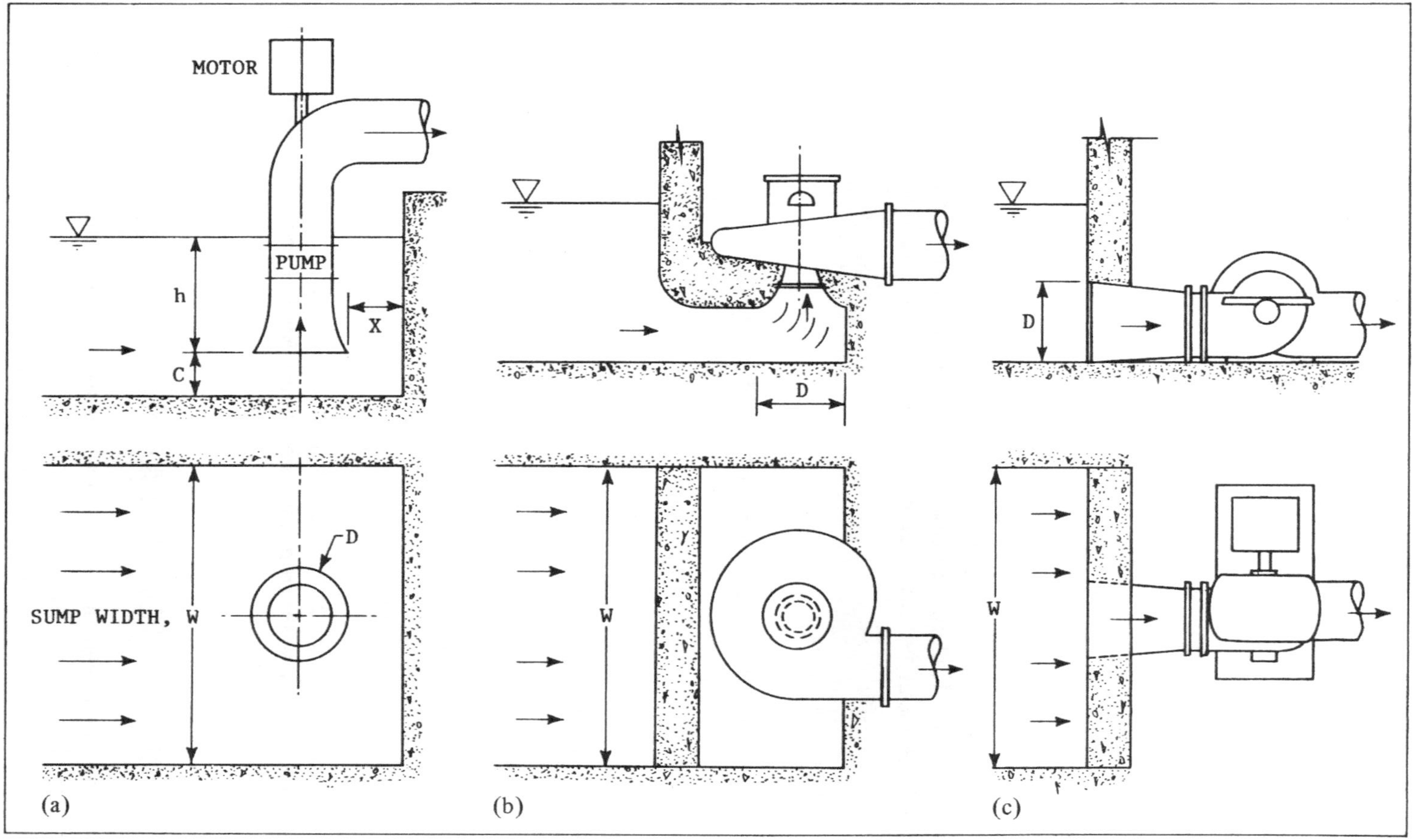

Figure 6.2.1. Sumps with wet and dry well: (a) wet well, axial flow pump; (b) dry well, vertical flow centrifugal pump; (c) dry well, horizontal flow centrifugal pump.

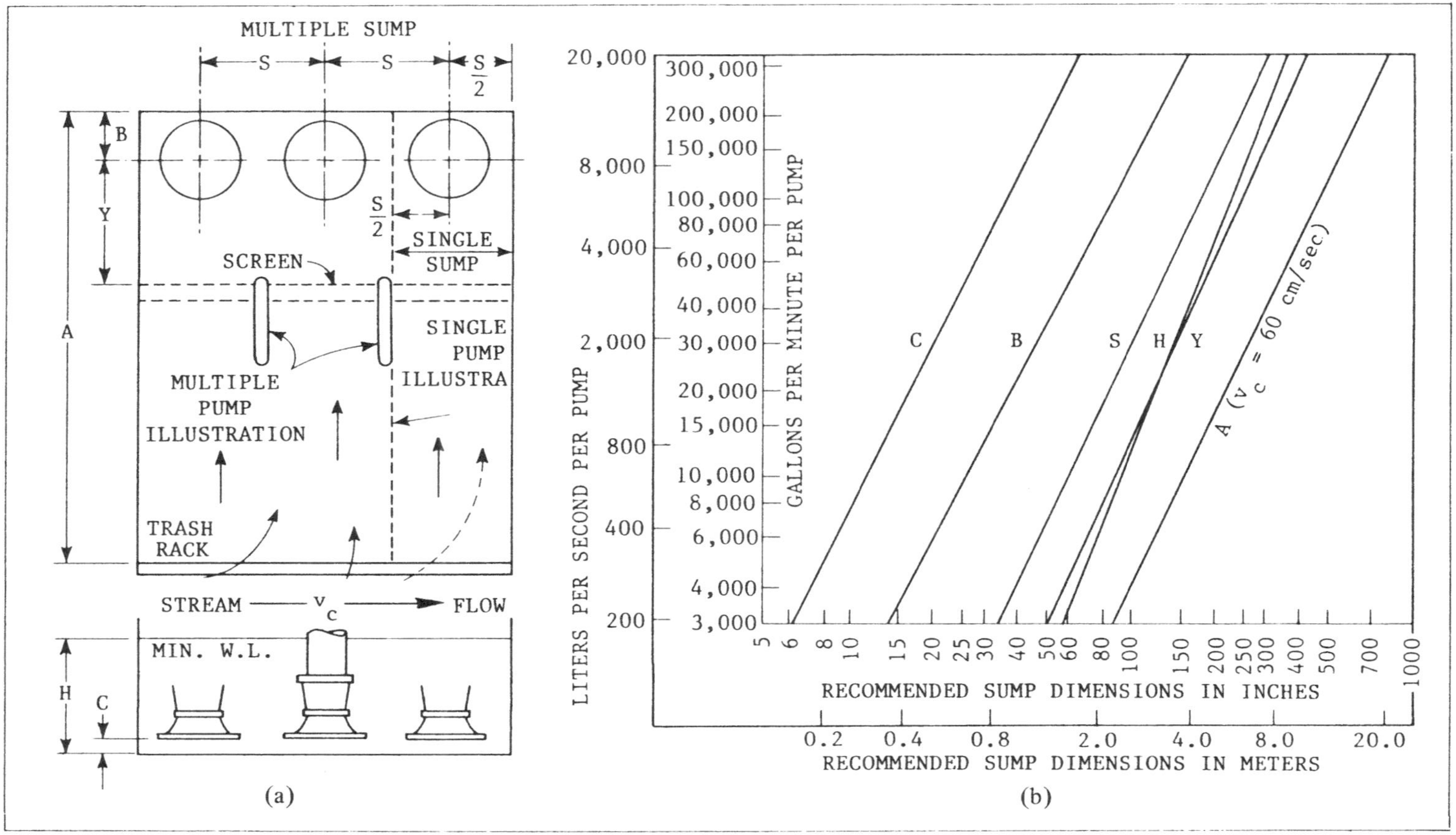

Figure 6.2.2. Recommended sump dimensions from the Hydraulic Institute Standards (1975): (a) definition of sump dimensions, (b) sump dimension versus flow.

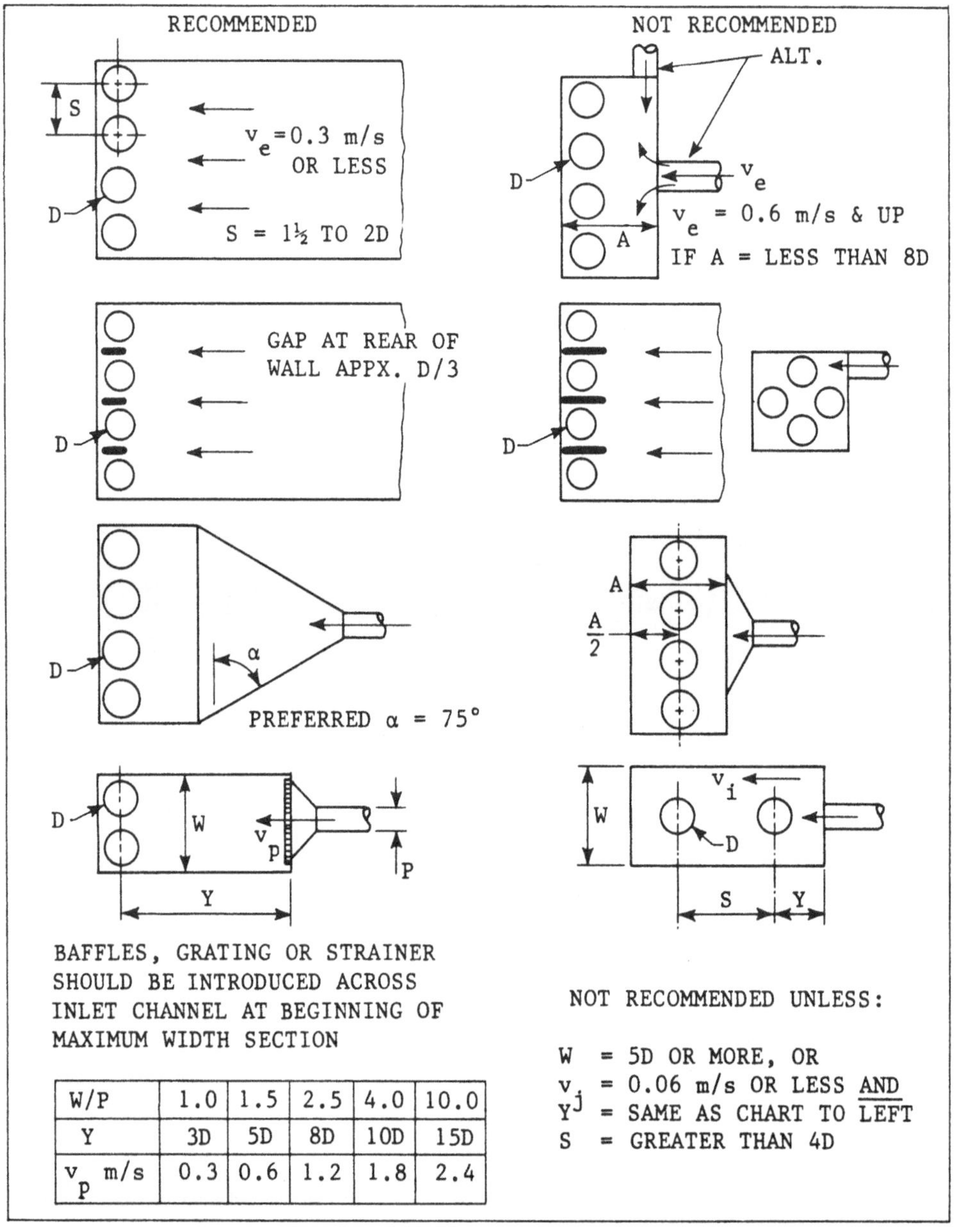

W/P	1.0	1.5	2.5	4.0	10.0
Y	3D	5D	8D	10D	15D
v_p m/s	0.3	0.6	1.2	1.8	2.4

Figure 6.2.3. Guidelines for multiple pump sumps (Hydraulic Institute Standards, 1975).

publication by Prosser (1977). However, these guidelines are mainly intended for simple straight approach intakes with reasonably uniform approach flow conditions to the sump. For complex situations involving non-straight approaches, multiple-bay intakes with many modes of operation, and for large capacity pumps, an evaluation of the preliminary design and derivation of modifications to the design can be accomplished only through a hydraulic model study. The recommended

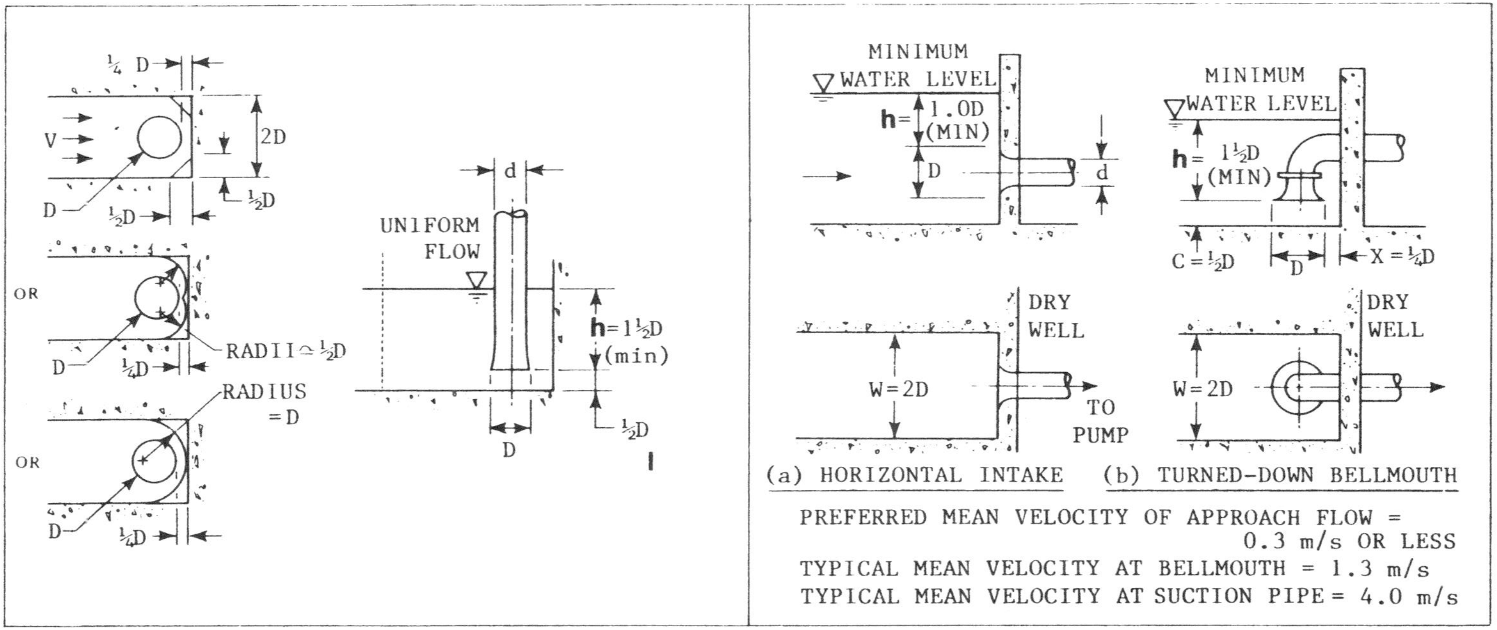

(a) (b)

Figure 6.2.4. Basic sump designs for single pump sumps (Prosser, 1977).

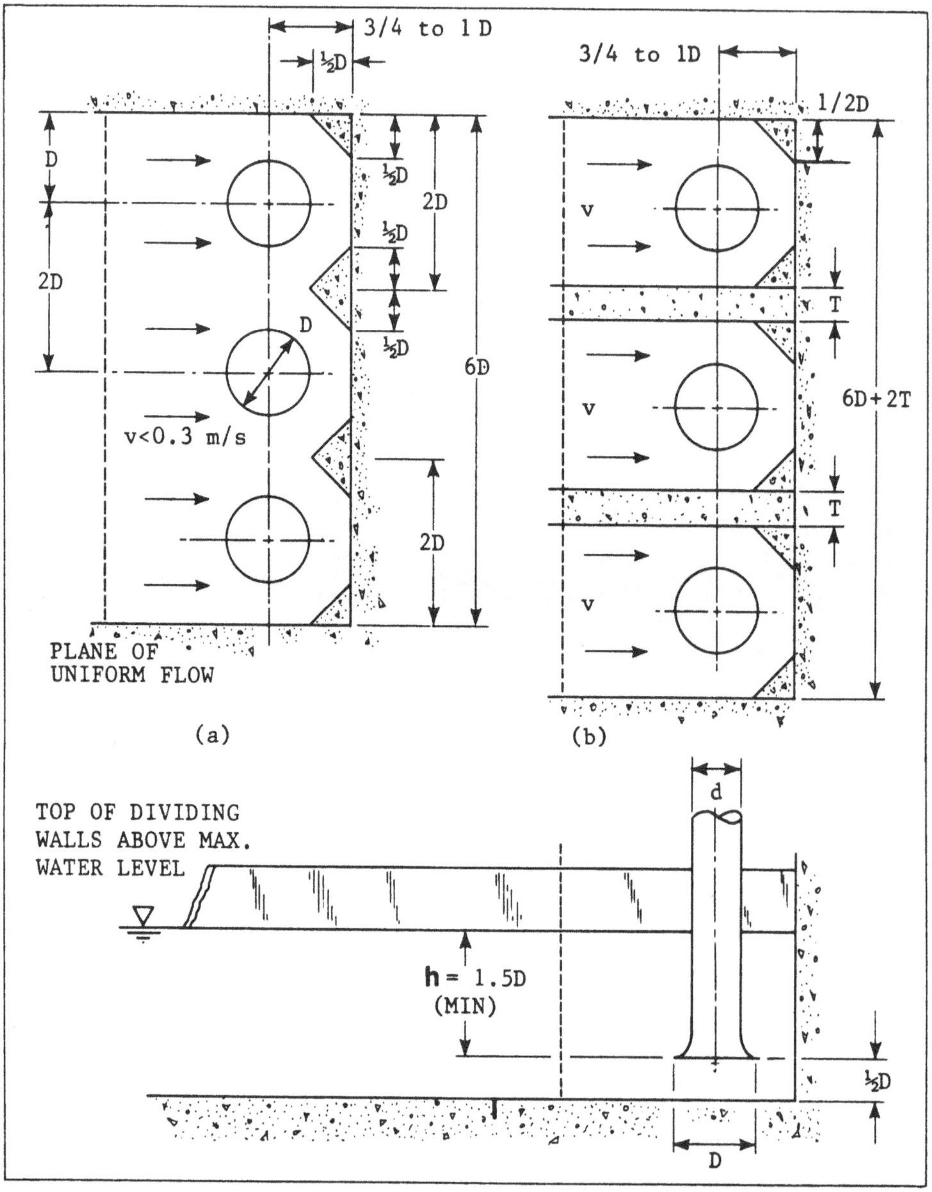

Figure 6.2.5. Basic sump designs for multiple pumps, wet well arrangement: (a) open sump, (b) unitised sump (Prosser, 1977).

sump dimensions for various pump flows as given in the Hydraulic Institute Standards (1975), are shown in Figures 6.2.2 and 6.2.3. The criteria used in arriving at the recommendations are not clear; but presumably are based on experience. Also, it appears that the guidelines are mainly written for sumps with vertical upward suction type outlets. However, the major parameters, such as wall clearances and minimum required submergences are, in general, found not to vary significantly for horizontal or vertical downward suction type outlets under the same flow conditions (Padmanabhan, 1982a). Relating the sump dimensions to flow only rather than flow and pump suction bell diameter may be confusing since vortexing is known to be affected by the velocity at the bell entrance for a given submergence. An approximate relationship between design flow and required pump suction bell diameter, may be obtained by assuming a velocity at the suction bell (v_D) of about 2.4 m/s, a value usually used by many U.S. pump manufacturers.

In the BHRA publication by Prosser (1977), recommended designs for single pump wet well and dry well sumps and basic sump designs for multiple pump bays are given and these are illustrated in Figures 6.2.4 and 6.2.5. The basis for the minimum submergence is the onset of air-entraining vortices. The minimum submergence is prescribed in terms of pump suction bell diameter but the ranges of velocity at the bell entrance for which the prescribed values are applicable are not stated.

6.2.6. IMPORTANT SUMP PARAMETERS

The various sump parameters that could affect the pump performance due to their influence on free-surface and sub-surface vortexing in the sump are discussed below (see Figure 6.2.1 for notations):

6.2.6.1. *Approach channel*

The geometry and orientation of the approach channel with respect to the pump bay (or bays) and the velocity of flow in the channel are very important since they all affect the circulation inside the sump and thereby contribute to vortexing. Side wall expansions or floor slope changes in the approach channel near the pump bay usually cause flow separations and formation of eddies contributing to vortexing. Side wall expansions, if unavoidable, should have included expansion angles preferably less than 20°. Floor slope changes should be less than 10°. Similarly, an off-set or bend in the approach channel near the pump bay entrance would be undesirable since it would generate a non-uniform flow into the bay. Non-streamlined obstructions such as screen supports located at the sump entrance could also to a lesser extent contribute to flow separation and eddying. Intuitively, the lower the velocities of approach flow, the lower would be any circulation inside the sump due to approach flow non-uniformities. Generally an approach flow velocity of less than 0.3 m/s is considered acceptable.

6.2.6.2. *Minimum desirable submergence at pump suction bell, h*

The minimum desirable submergence at the pump suction bell should not be less than a critical value below which objectionable surface vortices would occur. Unfortunately, objectionable vortices and hence the critical submergence are defined differently by individual researchers. The most common definition of an objectionable

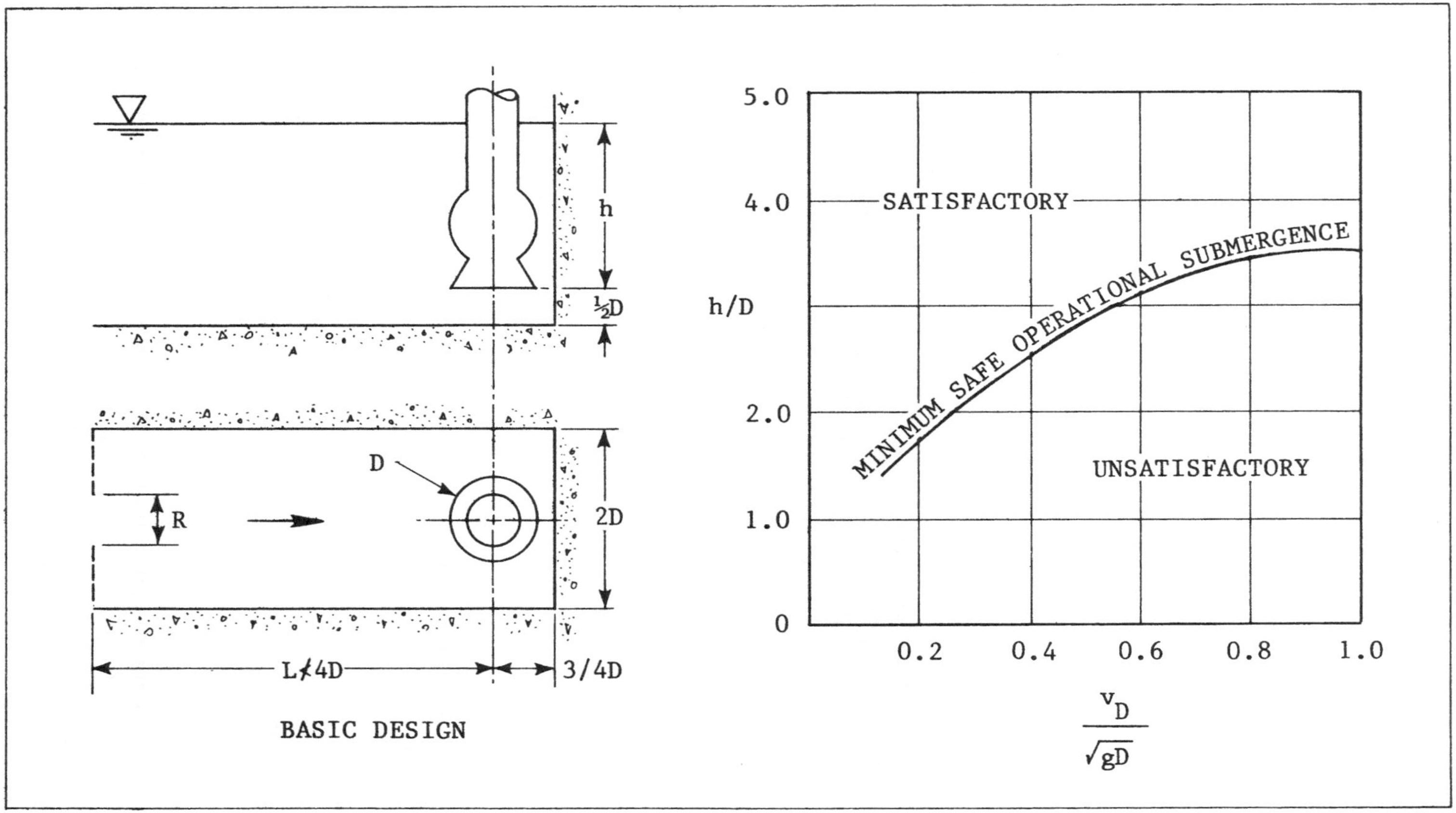

Figure 6.2.6. Recommended submergences for a basic design (Paterson & Noble, 1982).

vortex is one that is on the verge of drawing air into the suction pipe. The critical submergence is dependent on the approach flow patterns and other sources of vorticity existing in the sump and hence a universal value of critical submergence is not meaningful. The values of critical submergence given in design guides such as Hydraulic Institute Standards (1975) or BHRA publication by Prosser (1977) cannot be considered generally applicable, since the critical submergences are prescribed with no reference to velocity in the suction bell nor values of circulation.

Based on field experience as well as hydraulic model studies of both single and multiple pump sumps, Peterson & Noble (1982) have suggested a minimum safe operational submergence for a range of Froude numbers

$$Fr = \frac{v_D}{\sqrt{g\,D}}$$

as shown in Figure 6.2.6. The curve of Figure 6.2.6 is applicable if the approach flow to the sump is reasonably uniform and is more meanginful since it included the suction pipe velocity. A margin of safety was incorporated in deriving the curve. However, the curve is limited to Froude numbers of less than 1.0 with no data to support a flattening trend at higher Froude numbers. Instead of a curve, a straightline

$$h/D = 1.5 + 2.5\ Fr$$

would be a more appropriate envelope of the data to indicate an increasing minimum operational submergence with increasing Fr.

In general, the minimum required submergence could be expressed as

$$h = a + b\ Fr^n$$

where

a and b are constants which should depend upon the pump sump geometry.

Knauss has proposed a value of a = 1/2, b = 2 and n = 1 in Chapter 4 while Hecker has suggested a value of a = 1, b = 2.3 and n = 1 in Chapter 8. Based on a straight line (n = 1) to envelope the curve proposed by Paterson & Noble, a and b will have a value of 1.5 and 2.5 respectively. A value of a of at least 1 is appropriate so that at low Froude numbers a submergence of more than 1D would be available.

Experiments by Padmanabhan (1982b) using horizontal outlets indicated that air-core vortices are less likely if the submergence Froude number Fh is kept below about 0.6 for pump sumps with uniform approach flows. For Fh, below about 0.25 even with non-uniform approach flows air-core vortices are unlikely (Padmanabhan, 1982a; Pennino & Hecker, 1979).

6.2.6.3. *Floor clearance, C*

Floor clearance affects the available submergence since the higher the floor clearance the lower will be the available submergence on the suction bell entrance for a given water level. Hence, it is important not to provide a more than necessary floor clearance. But, very low floor clearances may enhance formation of submerged vortices. Commonly recommended values of the floor clearance lie between 0.4 to 0.75 bell diameter (Prosser, 1977; Paterson & Noble, 1982; Chang, 1977) and often the pump manufacturer's input is required to select a suitable value in this range.

6.2.6.4. *Back wall clearance, X, and bay width, W*

The width of the pump bay governs the approach velocity while in the case of wet well pump sumps, the back and side wall clearances from the outer periphery of the bell would affect the frequency and strength of any vortices. Very small backwall clearance gives rise to submerged vortices. A bay width of not less than 2 bell diameters and a back wall clearance of not less than 0.25 bell diameter are commonly recommended (Prosser, 1977; Paterson & Noble, 1982; Chang, 1977). The pump column is usually located in the mid-width of the bay which will fix the side wall clearances. A solid boundary interacts with a vortex making the vortex unstable and it is advantageous to have the back wall clearance small. An upper limit of 0.5 bell diameter is often recommended (Prosser, 1977).

6.2.7. RECOMMENDED BASIC DESIGN AND DESIGN CORRECTIONS

Combining available information on designs of pump sumps and using the experience from model studies conducted at the Alden Research Laboratory, recommended basic designs for a single bay and a multiple bay sump, assuming a reasonably uniform approach flow to the sump, are illustrated in Figures 6.2.7 and 6.2.8. For the required minimum submergences a straight line relationship

$$h/D = a + b\ Fr \text{ (relating the submergence to a Froude number)}$$

is recommended.

Suggested ranges of values of a and b are a = 1 to 1.5 and b = 2 to 2.5. It may not always be possible to provide a uniform approach flow to a sump due to practical limitations. When any problems are apparent in a sump design, corrective revisions are possible. It is recommended that such revisions be tested for their effectiveness in a hydraulic model. Most commonly used corrective measures drawn from available publications are given in the following paragraphs:

6.2.7.1. *Approach flow*

Approach flow corrections usually involve one or more of: (a) reducing the approach velocities by spreading the inflow; (b) eliminating flow separation by filling in any sharp corners by smooth transitions or by guide walls; (c) introduction of head loss devices such as weirs, surface beams, and curtain walls to improve flow pattern to the sump; and (d) relocation of pump positions in multiple pump sumps. Examples are given in Figure 6.2.9.

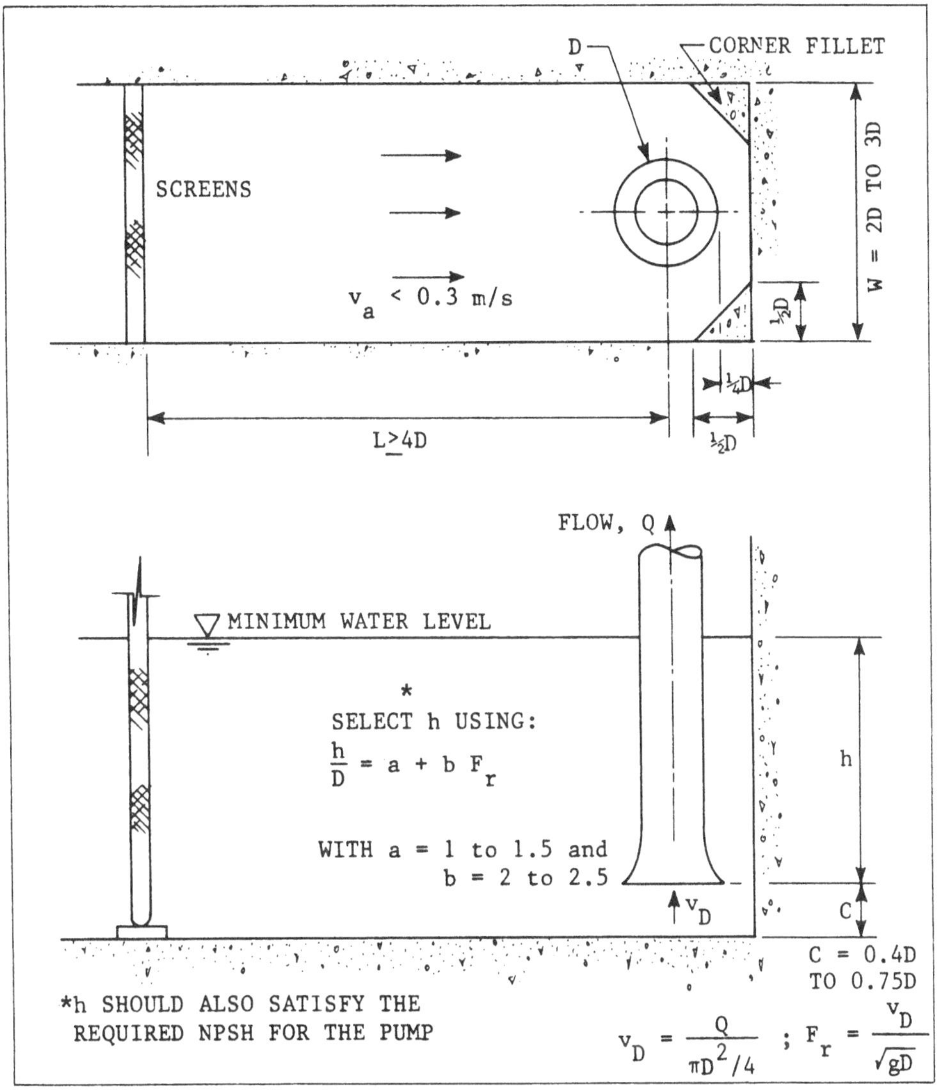

Figure 6.2.7. Basic design for single bay sump with uniform approach flow.

6.2.7.2. *Minimum submergence*

When the approach flow non-uniformities cannot be corrected, the submergence of the suction bell from the water surface may be increased sufficiently to avoid formation of free-surface air-core vortices. Even under highly non-uniform approach flows, submergences that would give a submergence Froude number, Fh of less than 0.25 were found sufficient to avoid air-drawing vortices in pump sumps based on experimental studies of Padmanabhan (1982a) as well as field evaluations of Pennino & Hecker (1979).

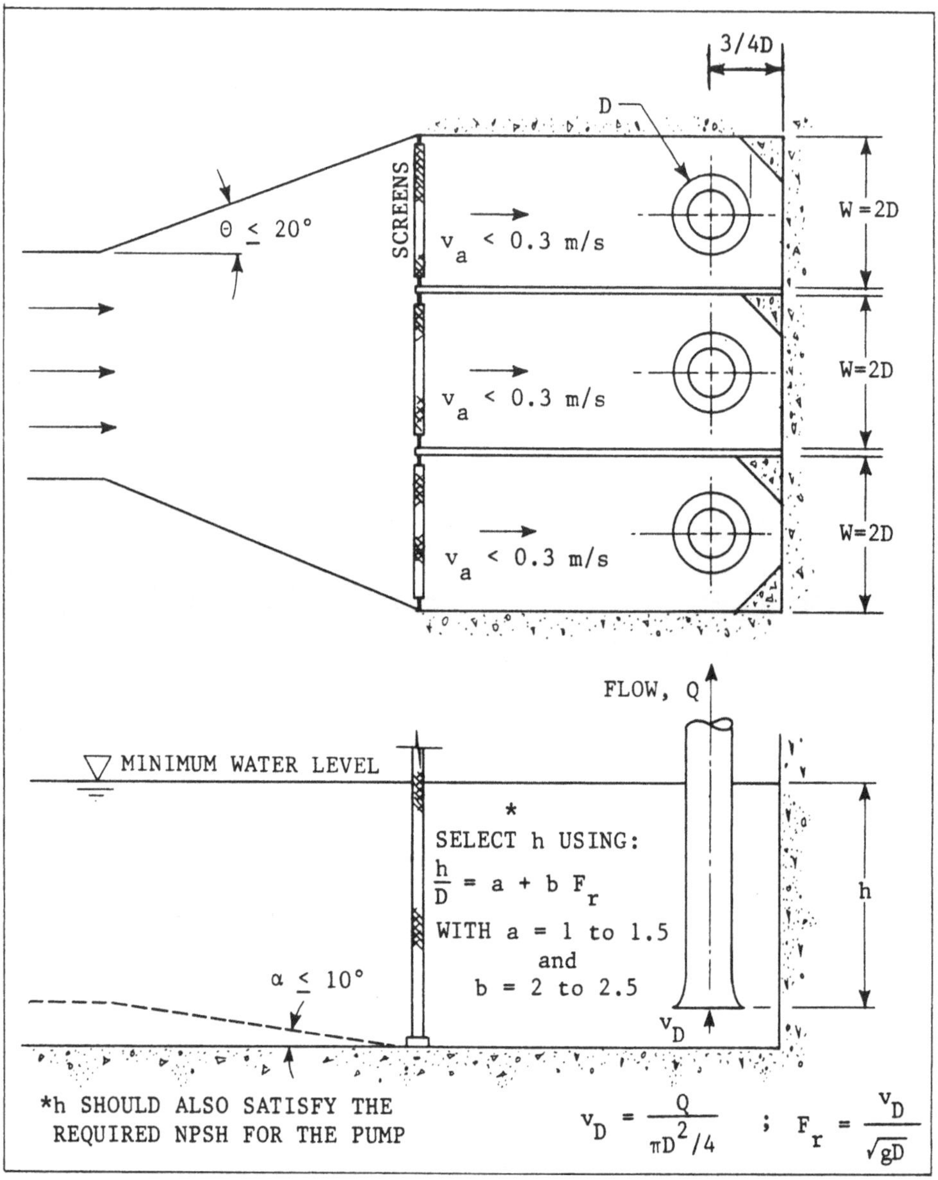

Figure 6.2.8. Basic design for multiple bay sump with uniform approach flow.

6.2.7.3. *Vortex suppressors*

A most effective free-surface vortex suppressing device is a horizontal floor grating installed about 10 to 15 cm below the water level at which strong vortexing is expected (usually the minimum water level). The depth and spacing of bars in the grating are important parameters. Usually bar depths of about 4 to 6 cm would be sufficient with a bar spacing of about 2.5 cm. Grating cage is also

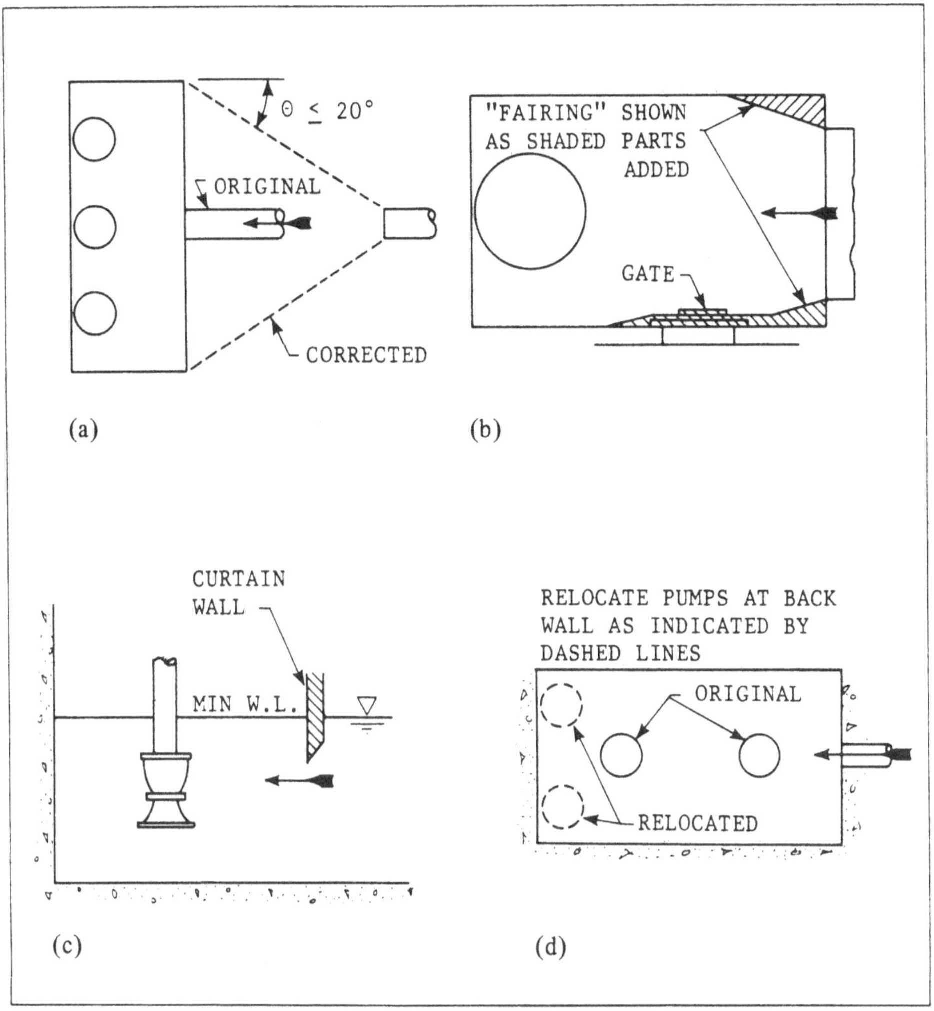

Figure 6.2.9. Typical approach flow corrections: (a) spreading flow, (b) filling in corners, (c) use of curtain walls, (d) relocation of pump position (mostly from Hydraulic Institute Standards, 1975).

effective, if it is large enough and is submerged just below minimum water level. Floating rafts are also used as vortex suppressors. In cases where the approach flow nonuniformities contribute to strong vortex formation a curtain wall extending across and into the water surface in the approach channel is found to be a very effective vortex suppressor. The curtain wall should be of sufficient depth usually obtained by trial in a model study. Examples of typical vortex suppressor arrangements are given in Figure 6.2.10. Subsurface vortices are closely linked with the flow patterns near the suction bell and are eliminated by one or more of the corrections described in the next item.

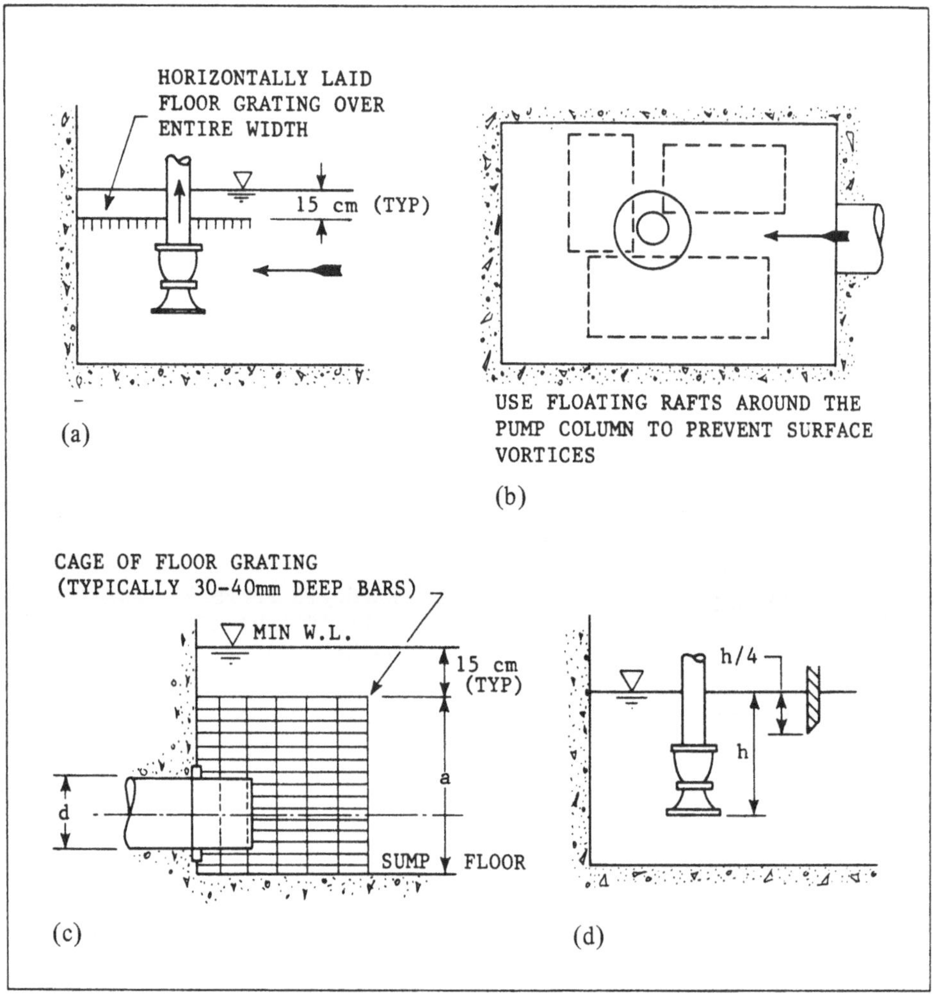

Figure 6.2.10. Typical Free-surface vortex suppression devices: (a) horizontal floor grating, (b) floating rafts, (c) grating cage, (d) curtain walls (Padmanabhan, 1982a; Hydraulic Institute Standards, 1975; Pennino & Larsen, 1982).

6.2.7.4. *Flow at or near suction bell*

Vortexing, swirl, pre-rotation, and uneven flow distribution to impeller are all influenced by the flow circulation at or near the suction bell, the flow pattern being affected by floor and wall clearances. Corrections to reduce this circulation are based on reducing rotational flow patterns and avoiding abrupt flow separations by changes in wall or floor clearances, floor splitters, wall splitters, turning vanes, and floor cones. Examples are illustrated in Figure 6.2.11.

6.2.7.5. *Pipe flow swirl*

Swirl in the suction pipe flow is normally reduced by all of the above corrective

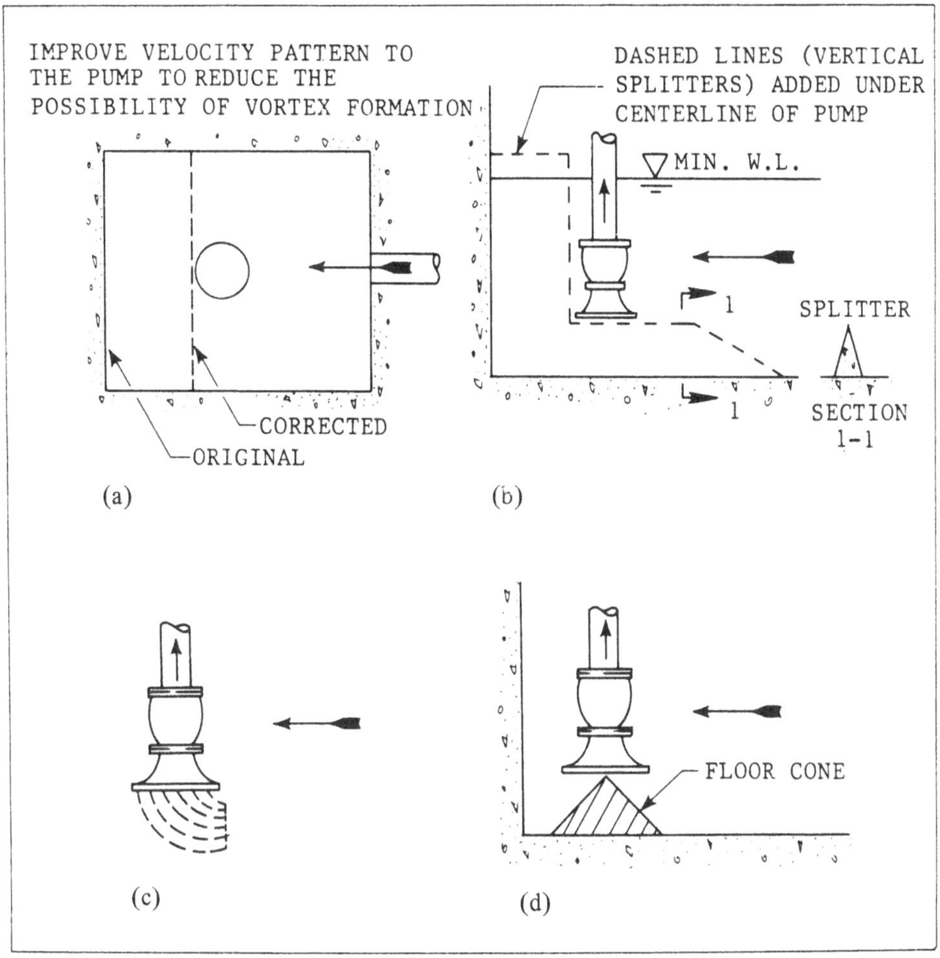

Figure 6.2.11. Typical sub-surface vortex suppression devices to correct flow at or near bell: (a) change of wall clearance, (b) floor and wall splitter, (c) turning vanes, (d) floor cone (mostly drawn from the Hydraulic Institute Standards, 1975).

measures designed to improve flow patterns approaching the sump or near the suction bell and to suppress vortices. However, as an additional measure, splitter vanes or cruciforms with 4 to 8 vanes may be added in the suction bell. The vanes can be of solid plate or perforated plate. Figure 6.2.12 illustrates a vane or curciform arrangement used in an installation.

6.2.8. HYDRAULIC MODEL-AIDED-DESIGN

Typically, a hydraulic model-aided-design is recommended for the following cases:

a) Pump sumps with non-symmetric approach flow (such as due to an offset

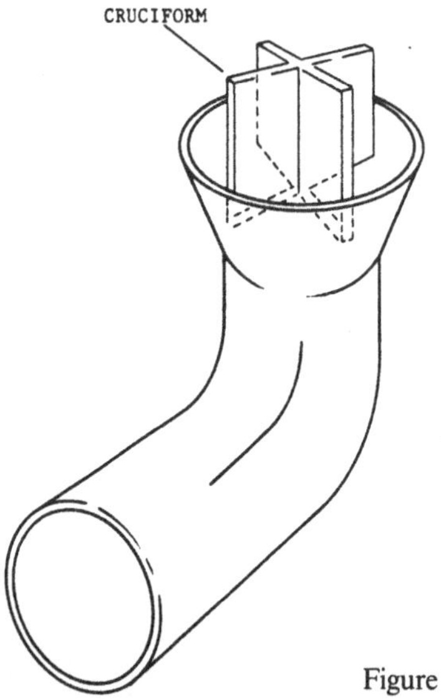

Figure 6.2.12. A cruciform swirl reducing device (Elder et al., 1978).

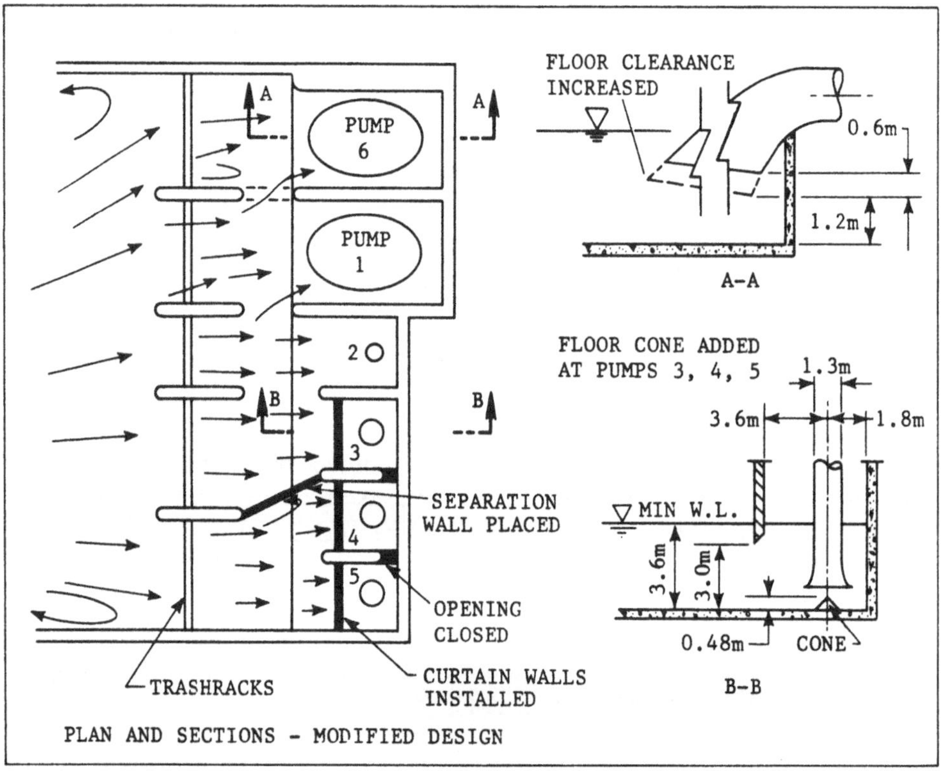

Figure 6.2.13. Hydraulic model-aided design modifications for a multiple pump station (Pennino & Larsen, 1982; 1:17 scale model; courtesy of Alden Research Laboratory).

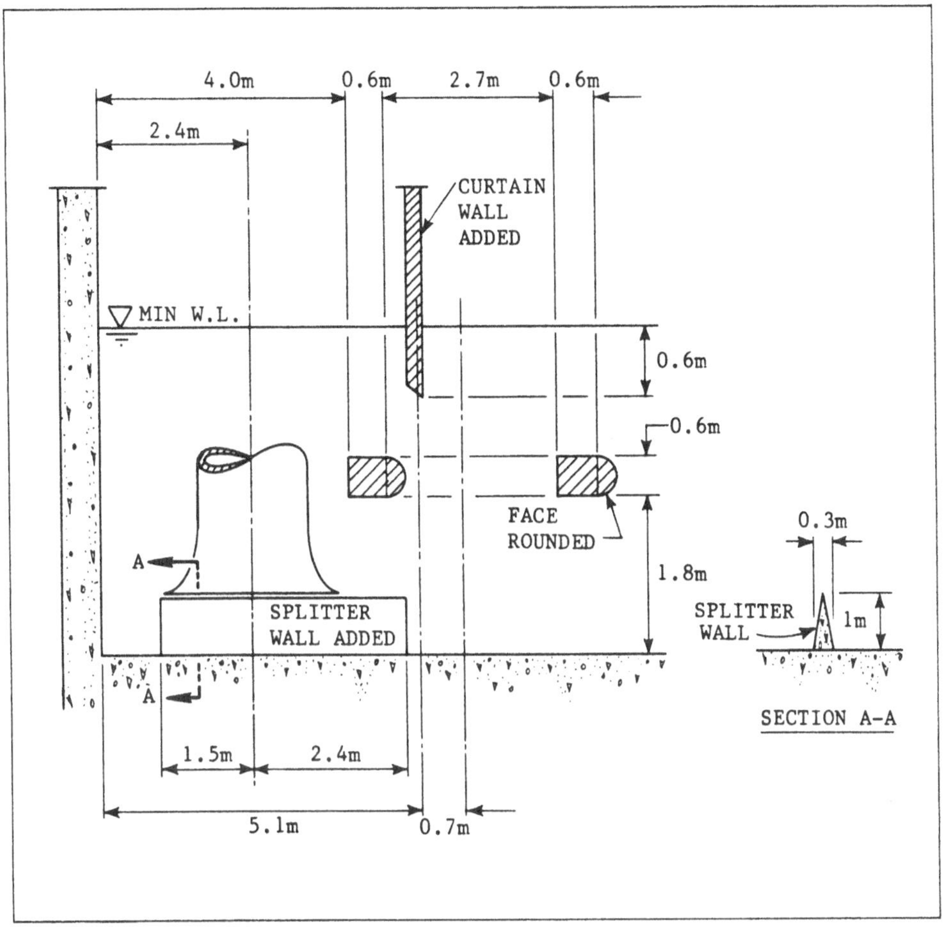

Figure 6.2.14. Hydraulic model-aided design revision of a pump sump (White & Larsen, 1979; 1:9 scale model; courtesy of Alden Research Laboratory).

or bend in the approach channel upstream near the sump) if the submergence Froude number

$$Fr = \frac{v_D}{\sqrt{g\,h}} > 0.25$$

b) Multiple pumps with a common approach channel and a multitude of pump operating sequences possible.

c) Pump sumps of large capacities greater than about 150 m^3/min. per pump (considering the cost of such large installations, a model study is warranted).

d) Pump sumps with possibilities of obstructions to approach flow close to the pump location, such as debris blockage of trash racks or screens.

e) Pump sumps with minimum water levels below the recommended values, but practical limitations do not allow a higher water level.

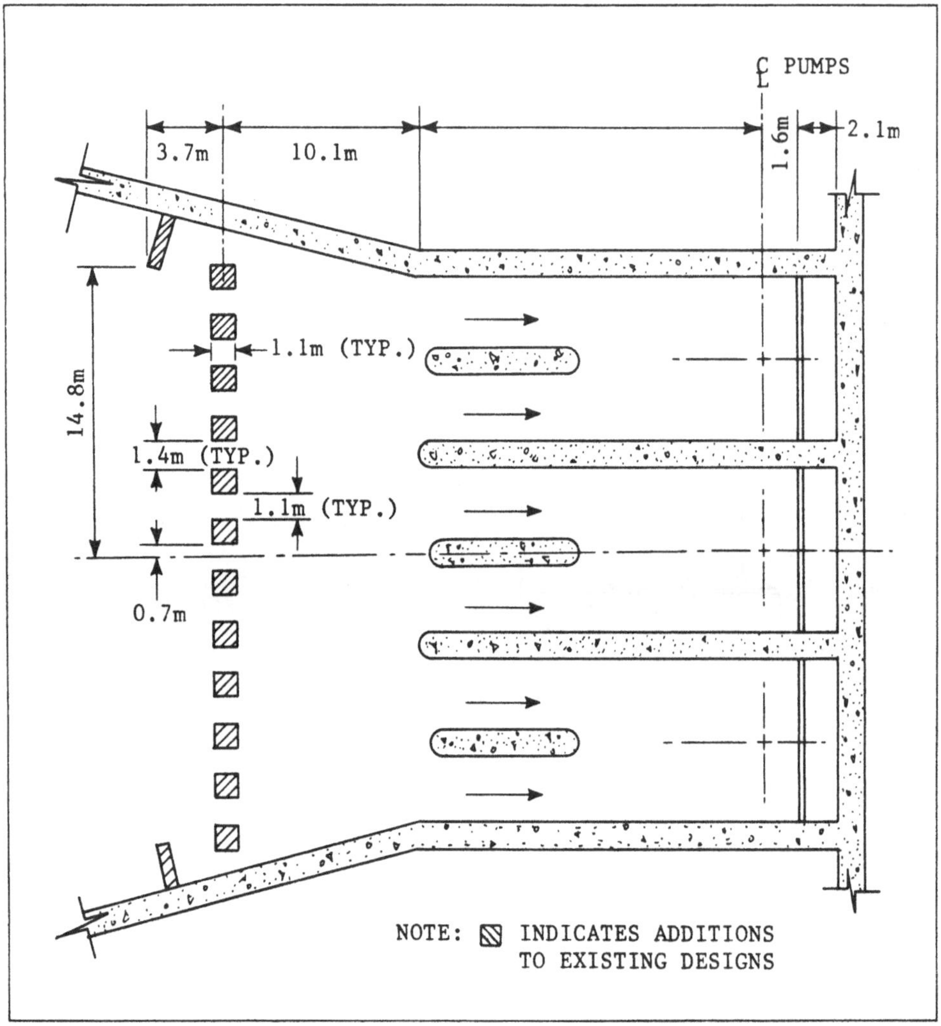

Figure 6.2.15. Pump bay modifications using a 1:9 scale hydraulic model (Bozoian & Larsen, 1977; courtesy of Alden Research Laboratory).

6.2.9. EXAMPLES OF HYDRAULIC MODEL-AIDED-DESIGN CORRECTIONS

Examples of hydraulic model-aided-design corrections based on hydraulic model studies conducted at the Alden Research Laboratory are illustrated in Figures 6.2.13 to 6.2.16. Corresponding to each of the figures, details of the original design, problems found by model tests and the fixes derived to rectify the problems are listed in Table 1. Usually the corrections involve adopting more than one of the corrective measures described in the previous paragraphs. These design corrections,

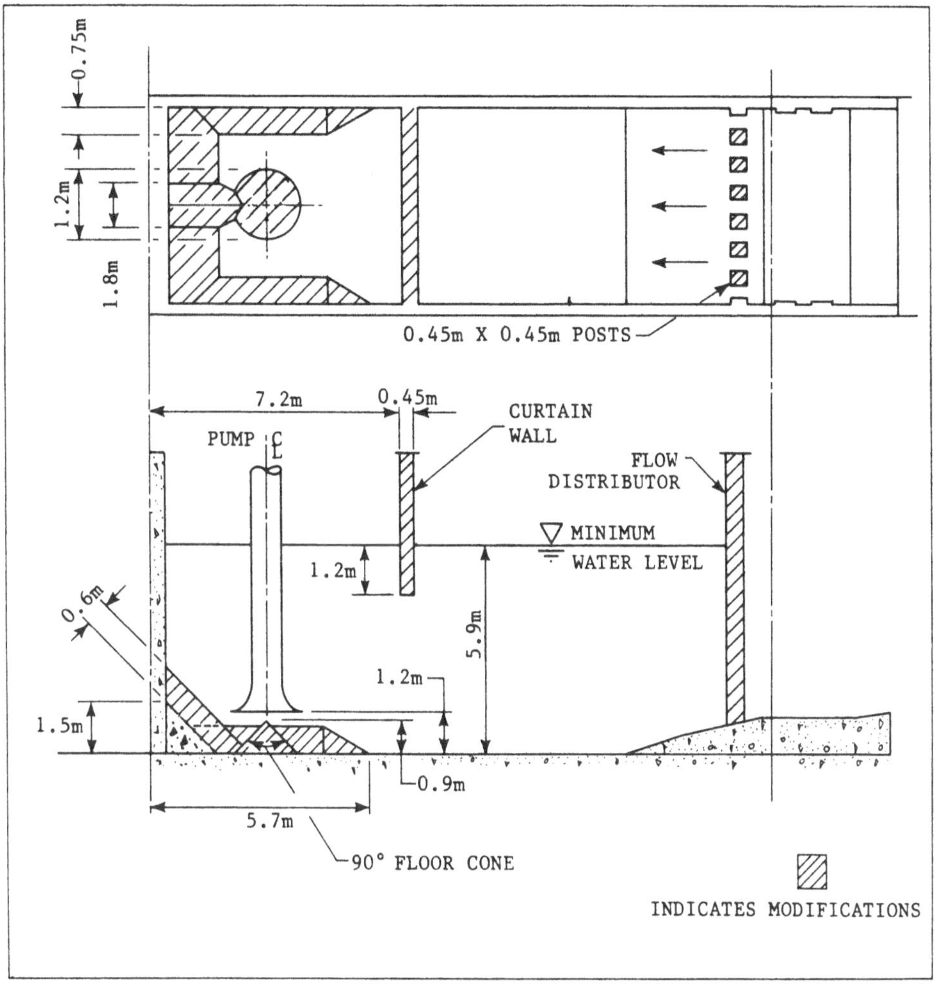

Figure 6.2.16. Final design of a sump derived from a model study (White & Larsen, 1983; 1:10 scale model; courtesy of Alden Research Laboratory).

especially the geometric dimensions associated with them, are specific for the particular sump and without a hydraulic model study such corrections can seldom be derived with confidence.

6.2.10. SUMMARY

The sources of adverse problems in pump operations attributable to the sump design usually stem from one or more of the following: unsatisfactory approach flow, insufficient submergence, and inadequate floor and wall clearances at the suction bell. Available guidelines including those given in here may be used for

Table 6.2.1. Details of case studies shown in Figures 6.2.13 to 6.2.16 (prototype vales for dimensional variables).

Case study	Q m^3/s per pump	D m	V_D m/s	h/d	x/D	c/D	Fr	Problems encountered in model	Fixes used in model
Fig. 6.2.13 Pumps 3, 4, 5	8.5	2.3	2.0	1.5	0.3	0.4	0.42	Air-drawing free-surface vortices; floor vortices, swirl angles in suction pipe above 10°	Curtain wall to suppress free-surface vortices; floor cone to eliminate floor vortices and reduce swirl; also, extension of bay walls and placing of a separation wall
Fig. 6.2.14	18.0	2.9	2.7	1.3	0.4	0.3	0.51	Air-drawing free-surface vortices, floor vortices	Curtain wall to suppress free-surface vortices; rounding of beam corners splitter wall to eliminate floor vortex
Fig. 6.2.15	14.0	2.9	2.1	1.7	0.7	0.5	0.39	Air-pulling free-surface vortices due to non-uniform approach flow	Flow distributor piers at the entrance channel to eliminate free-surface vortices
Fig. 6.2.16	9.0	2.6	1.7	2.0	0.7	0.5	0.34	Trash-pulling vortices; high pre-rotation	Curtain wall to suppress free-surface vortices; splitter wall and floor cones to reduce prerotation

developing a basic sump design; but these guidelines are mainly based on a reasonably uniform approach flow. Due to practical limitations, the approach flow may be far from uniform and, in such cases, the best way of arriving at a most satisfactory design, ossuring a trouble-free pump operation, would be to start from a basic design and then resort to a hydraulic model-aided verification and correction to the basic design. In deciding the corrections, the several examples of model aided corrections illustrated herein would be helpful. A joint involvement of the pump manufacturer and the sump designer should be considered important in evolving a final design.

A hydraulic model-aided design is also considered necessary for sumps with large pumps of flows above 150 m^3/min., for multiple pump sumps with a common approach channel if operation of several pump combinations is possible, and for pump sumps involving significant obstructions at sump entrance such as debris blockage of sump screens.

Vortex-flow intakes 7

S.C.JAIN & R.ETTEMA
University of Iowa, Iowa City, USA

7.1. INTRODUCTION

Whereas in other chapters of this volume, swirling flows are presented as being undesirable for intakes, especially those connected to hydraulic machinery, this chapter deals with the design of intakes to produce controlled swirling flows.

It i often hydraulically efficient to convey water as a swirling flow down a vertical or near-vertical dropshaft. This requires that the approach flow enters a vortex-flow intake at the top of the dropshaft and then swirls down the dropshaft. The vortex-flow intake imparts an angular motion to the flow which then enters the dropshaft as a swirling annular jet with an air core at the center of the dropshaft, as shown in Figure 7.1. The air-core area initially decreases and then increases along the dropshaft, thereby forming a throat. As the flow passes down the dropshaft, its vertical velocity component increases, the swirl attenuates, and the flow direction gradually approaches the vertical. The flow, however, continues to hug the dropshaft's wall, even for a relatively low flow with a small tangential velocity component.

Certain characteristics of swirling flows in conduits make vortex-flow intakes appropriate for dropshafts. The spiralling flow along a dropshaft wall can result in considerable dissipation of the flow energy (Jeanpierre & Lachal, 1966; Jain & Kennedy, 1983). Further, less air is entrained by a coherent, wall-attached spiralling flow than by a plunge flow (Laushey, 1953; Whillock & Thorn, 1973, and others), and the entrained air tends to be squeezed back toward the air core by the outward pressure gradient produced by the rotation of the flow. For a more detailed account of dropshaft flows the reader is referred to Kleinschroth (1980), and Jain & Kennedy (1983).

7.2. CONFIGURATIONS OF VORTEX-FLOW INTAKES

Intake configuration controls both the flow depth in the approach channel and the dropshaft diameter required to pass the design flow with a minimum permissible

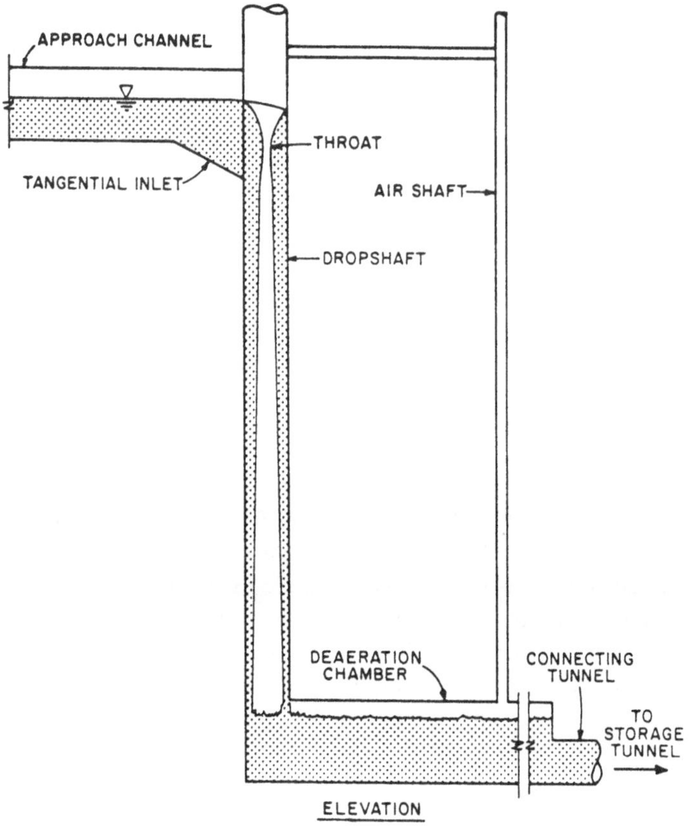

Figure 7.1. Vortex-flow drop structure.

air-core area. Several intake configurations are used to produce swirling flows, and these can be classified into five broad categories:

a) Circular: The side of the intake is concentric with the dropshaft and it has a horizontal floor (Figure 7.2a).

b) Scroll: The sides of the intake curl towards the dropshaft and it has a horizontal floor (Figure 7.2b).

c) Spiral: The sides of the intake curl toward the dropshaft and it has a sloping floor (Figure 7.2c).

d) Tangential: The sides of the intake are straight and make an eccentric approach to the dropshaft. The intake has a sloping floor (Figure 7.2d).

e) Siphonic: The intake consists of a series of siphons located around the entrance to a dropshaft (Figure 7.2e).

The development of vortex flow intakes is briefly reviewed in this section. Siphonic intakes are not discussed further here, as they are less commonly used due to their complex geometry.

Circular vortex-flow intakes were the earliest configuration studied and used (Laushey, 1952, 1953; Laushey & Mavis, 1953). They were put into service for

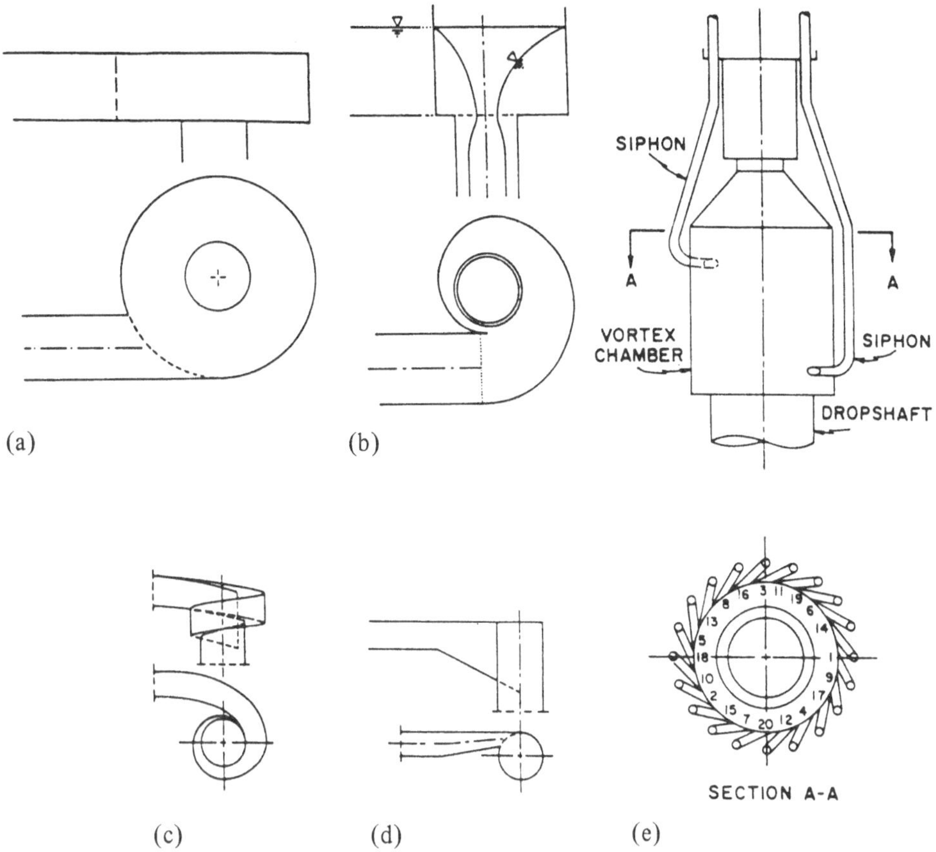

Figure 7.2. Vortex-flow intakes: (a) circular, (b) scroll, (c) spiral, (d) tangential, and (e) siphon.

discharges up to 2 m^3/s, in a storm-sewer system operated by the Allegheny County Sanitary District, Pittsburgh, USA. Later, when appraised (Anderson, 1961) for passing larger discharges (17 m^3/s and greater), circular intakes were discarded because their use required unacceptably large intake depths to discharge the requisite flows. An improvement to the hydraulic performance of circular inlets was proposed by Kleinschroth & Wirth (1981), who showed that inserting a guide wall (Figure 7.3) decreased the intake depth to pass a flow. Circular intakes which were pressurized by the use of a lid, have been studied by Giesecke & Brombach (1974), Hoffman & Widmar (1980) and Kleinschroth & Wirth (1981).

Circular intakes have been largely superseded by scroll and tangential intake configurations. The scroll configuration is possibly the most widely used vortex-flow intake, although it is likely (from the results presented by Jain & Kennedy, 1983) that it too will be superseded by the simpler and more compact tangential configuration. The hydraulic performance of scroll intakes has been studied by many researchers, including Viparelli (1950, 1961), Jeanpierre & Lachal (1966), Drioli (1969a, b), Kleinschroth & Wirth (1981), and Jain & Kennedy (1983). These

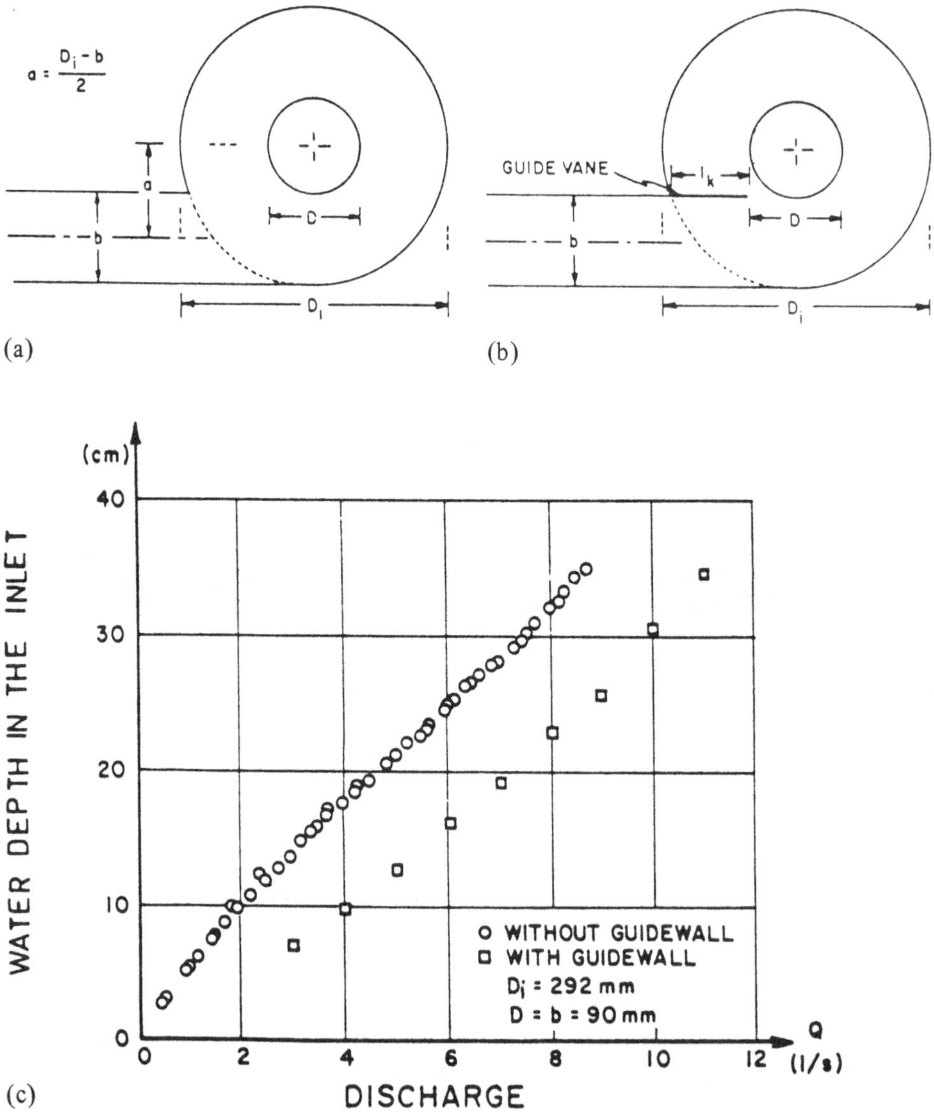

Figure 7.3(a) Circular inlet for a vortex-flow drop structure, (b) circular inlet with a guide vane, (c) comparison of head-discharge performances (Kleinschroth & Wirth, 1981).

studies show that scroll intakes require less head than do circular intakes.

Spiral intakes have been used in mountainous areas where the approach flow to a dropshaft can be conveniently supercritical. Gaspard et al. (1949), Jevdjevich & Levin (1953), Drioli (1969a, b), Vischer (1976), Desserich (1980), and Kleinschroth & Wirth (1980) have studied spiral intakes. Although appropriate at certain sites, the spiral intake is relatively complex in form. It too may be discarded in favor of the tangential intake configuration.

Tangential configurations are the simplest and most compact of the vortex-flow intakes. The approach flow channel is contracted and flared to the dropshaft so as to project the flow tangentially around the dropshaft wall. The hydraulic performance of tangential intakes has been studied by Jevdjevich & Levin (1953), Hausser & Galliere (1960), Brooks & Blackmer (1962), Slissky et al. (1979), and Jain & Kennedy (1983). Jevdjevich & Levin showed that the tangential and spiral configurations are equally effective in producing a controlled swirling flow. Jain & Kennedy (1983) have shown that the tangential configuration is as effective as the scroll configuration for producing a controlled swirling flow. The main attractive feature of a tangential configuration is its simple and compact form.

7.3. HYDRAULIC ANALYSIS

Flows through vortex-flow intakes have been analyzed by several investigators (Viparelli, 1950; Ackers & Crump, 1960; Knapp, 1960; Pica, 1970; Jain, 1984). Two, related hypotheses common to most analyses are that the throat (defined as the section where the air-core area is a minimum) is a control section of the flow, and that the principle of 'maximum discharge' is applicable at the throat. For circular, scroll and spiral intakes, the throat is assumed to coincide with the entrance of the dropshaft. Hence, the location of the control section is known and the discharge past the throat is equal to the total discharge through the intake. For a tangential intake, however, neither the location of the throat nor the discharge past the throat are known a priori. Hydraulic analysis are presented for scroll and tangential intakes only; the analysis for the scroll intake can easily be extended and modified for circular and spiral intakes

7.3.1. *Scroll intake*

The following analysis is based on the work by Pica (1969, 1970) who extended the Viparelli's solution (Viparelli, 1950), and on the following assumptions: (a) the energy head is a constant within the vortex-flow intake; (b) the pressure inside the air core is atmospheric; (c) the distribution of the tangential velocity component at the throat (Section 2 in Figure 7.4) is given by the free vortex relation

$$v_t r = c$$

where

v_t = tangential velocity component at a distance r from the dropshaft axis

c = constant.

(d) the pressure at all points in Section 2 is atmospheric; (e) both the water depth and velocity along the approach channel section (Section 1 in Figure 7.4) upstream from the vortex-flow intake remain practically constant.

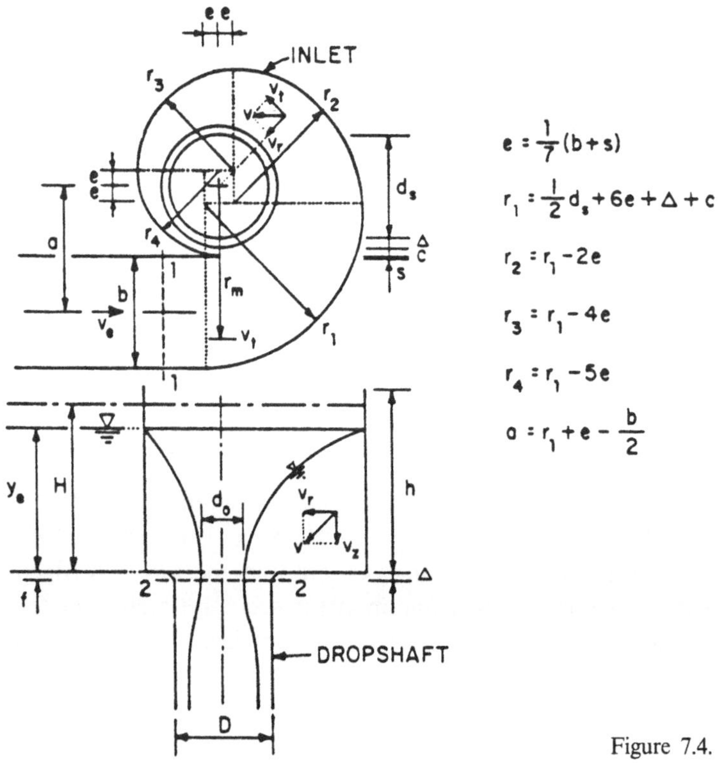

Figure 7.4. Standard scroll intake.

Application of the momentum and energy principles between Sections 1 and 2 yields:

$$v_e\, a = c \tag{7.1}$$

$$Q = C_\lambda \frac{\pi}{4} D^2 \sqrt{2g(H + \Delta)} \tag{7.2}$$

where

$$C_\lambda = \sqrt{1\text{-}E} - \lambda \sqrt{\lambda^2\text{-}E} - E\, \ell n \frac{1 + \sqrt{1\text{-}E}}{\lambda + \sqrt{\lambda^2\text{-}E}} \tag{7.3}$$

$$E = \frac{2c^2}{g(H + \Delta)D^2} \tag{7.4}$$

$$\lambda = d_o/D, \quad H = y_e + \frac{V_e^2}{2g}, \quad V_e = Q/by_e \tag{7.5}$$

and other symbols are defined in Figure 7.4. From the principle of 'maximum discharge' (involving minimum acceptable air-core area), $\delta Q/\delta\lambda = 0$, and one obtains

$$E = \lambda^2 \tag{7.6}$$

Inserting Equation (7.6) into both Equations (7.3) and (7.4) gives:

$$C_\lambda = \sqrt{1-\lambda^2} - \lambda^2 \ell n \frac{1+\sqrt{1-\lambda^2}}{\lambda} \tag{7.7}$$

and

$$\lambda = \frac{c}{D}\sqrt{\frac{2}{g(H+\Delta)}} \tag{7.8}$$

For a given discharge into a vortex-flow intake of known geometry there are three unknowns; namely, λ, y_e, and c. Conveniently, there are three Equations: (7.1), (7.2) and (7.6) to solve for them. Equations (7.1) and (7.2), after some algebraic manipulation, can be reduced to the following dimensionless forms which are more suitable for sizing an intake:

$$\frac{y_e}{D} = \frac{\pi}{2}\frac{a}{b}\frac{C_\lambda}{\lambda} \tag{7.9}$$

$$\frac{Q}{D^2\sqrt{g\,D}} = \frac{1}{2\sqrt{2}}\left[\frac{\frac{y_e}{D}+\frac{\Delta}{D}}{(\pi C_\lambda)^{-2} - \left(4\frac{b}{D}\frac{y_e}{D}\right)^{-2}}\right]^{\frac{1}{2}} \tag{7.10}$$

Comparisons of Equations (7.7.), (7.9) and (7.10) with experimental data are shown in Figures 7.5 and 7.6. Though, there is a good agreement between the theory and experimental data for the depth-discharge relation, the theory significantly over-predicts λ.

Pica (1970) also analyzed the flow for the condition v_t = constant. He found that the distribution of v_t has little effect on the depth-discharge relation, but has a significant effect on λ values. He recommended replacing Equation (7.7) with the following empirical relation:

$$C_\lambda = 0.95 - \lambda \tag{7.11}$$

For v_t = constant, Equation (7.9) can be written as

$$\frac{y_e}{D} = \frac{\pi}{2}\frac{a}{b}C_{\lambda o} \tag{7.12}$$

where

$$C_{\lambda o} = \frac{3}{2}\frac{(1-\lambda^2)^2}{(1-\lambda^3)}\frac{C_\lambda}{[1-\lambda^2)^2 - C_\lambda^2]^{\frac{1}{2}}} \tag{7.13}$$

Comparisons of Equations (7.10), (7.11) and (7.12) with experimental data are also presented in Figures 7.5 and 7.6. The theoretical and experimental values of λ are in good agreement when the theory includes the empirical relation Equation (7.11). For stable conditions in the vortex-flow intake, Pica further suggested that

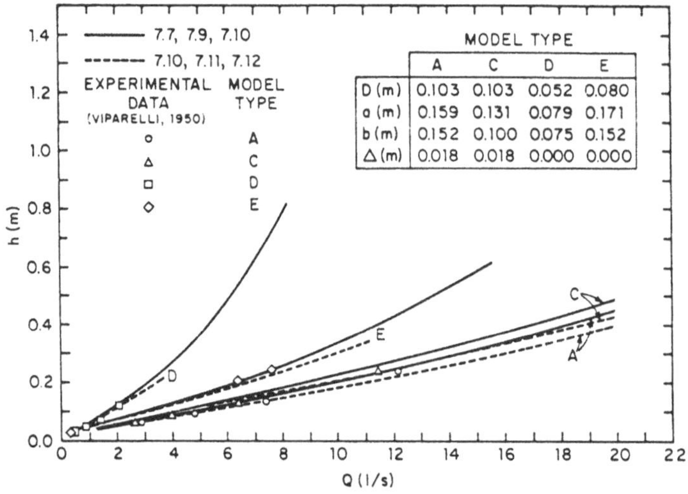

Figure 7.5. Comparison of depth-discharge curves.

$$\frac{a}{b} - 0.5 > 0.65\,\frac{D}{b} \tag{7.14}$$

and

$$\frac{\Delta}{D} = 0.1 \sim 0.2 \tag{7.15}$$

For sizing an intake, a suitable value for λ is selected in order to avoid choking of the throat). Viparelli (1950) recommended the use of $\lambda_{min} = 0.5$.

7.3.2. *Tangential intake*

The geometry of a tangential intake and the flow in it are shown in Figure 7.7. There exist three possible flow-control sections; one at the upstream end of the intake (Section 1 in Figure 7.7), the second at the downstream end (Section 2) of the intake, and the third at the throat (Section 3). In the following analysis (Jain, 1984), it is assumed that flow control at the throat does not affect significantly flow conditions at the second control section. The approach flow depth is then controlled by either Section 1 or 2.

7.3.2.1. *Depth-discharge relation*

For small flows, Section 1 acts as the control, so that flows in the approach channel and the tangential intake are critical and supercritical, respectively. For larger discharges, the first control is drowned by the second control which now governs the flow depth, so that flows in both the approach channel and the tangential intake are subcritical. The flow depth at the junction of the intake with the dropshaft is critical for large flow rates. The specific energy of the flow at the junction,

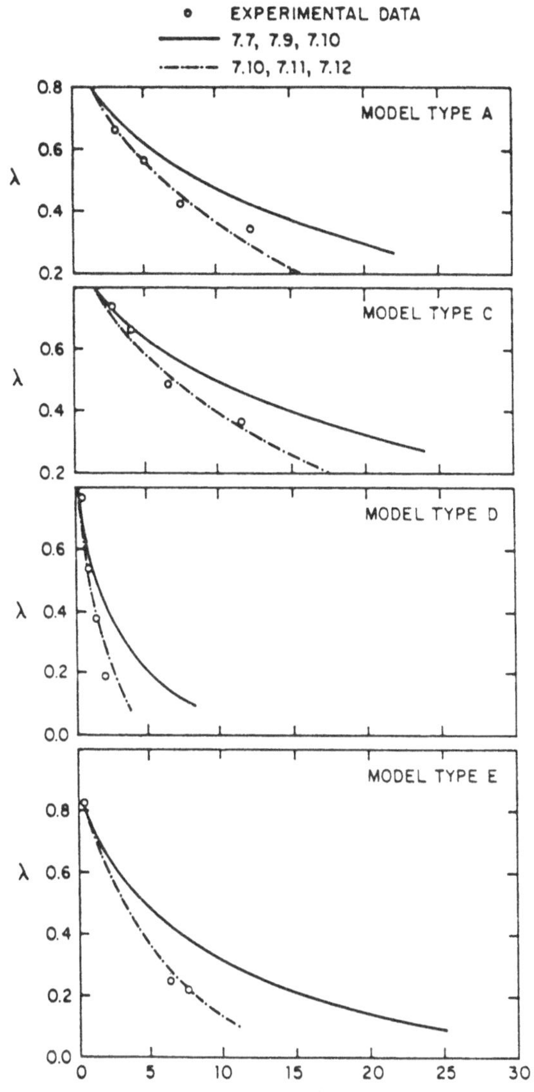

Figure 7.6. Comparison of air-core diameter.

E_j, is given by

$$E_j = \frac{3}{2}\left[\frac{\left(\frac{Q}{e}\right)^2}{g}\right]^{\frac{1}{3}} (\cos\beta)^{\frac{2}{3}} \tag{7.16}$$

where

Q = discharge
e = intake width at the junction
β = slope of the transition floor
g = gravitational acceleration.

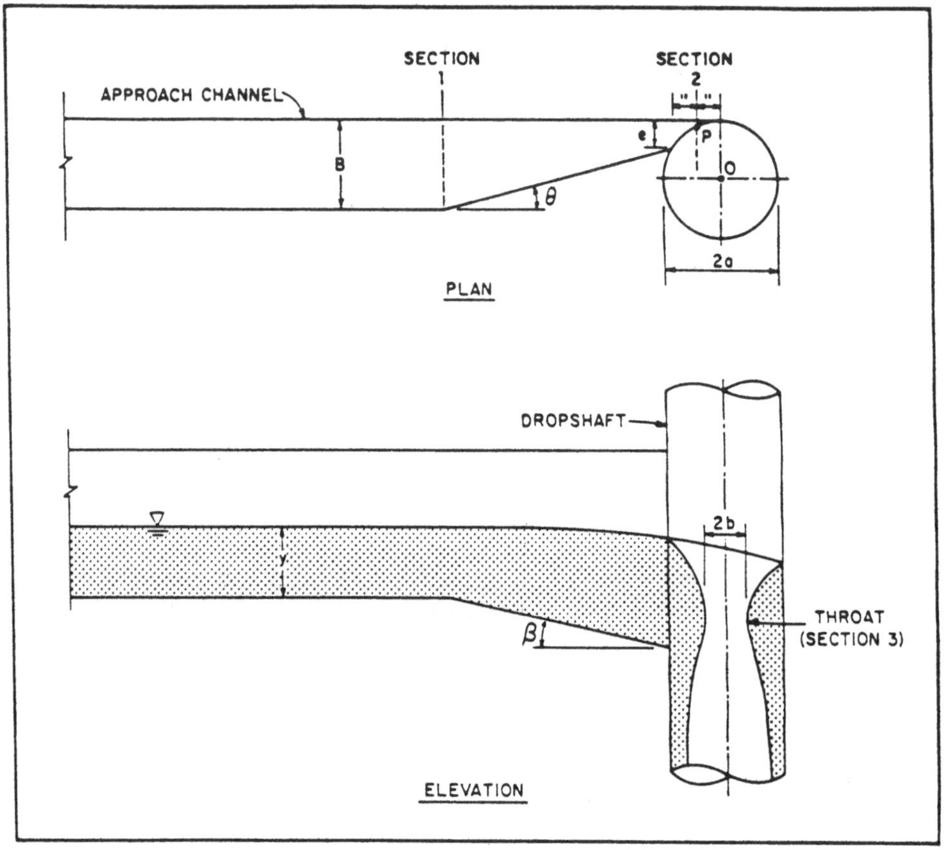

Figure 7.7. Tangential intake.

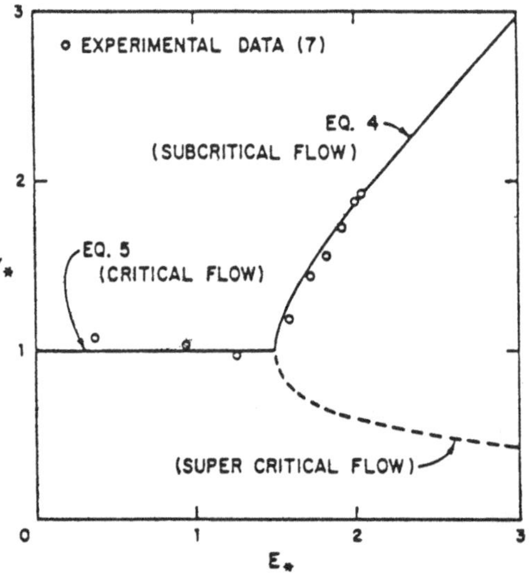

Figure 7.8. Comparison of depth-discharge curves.

If hydraulic losses in the intake are neglected, E_j can be related in a nondimensional form to the specific energy of the flow in the approach channel as

$$y_* + \frac{1}{2y_*^2} = E_* \qquad (7.17)$$

where

$y_* = y/y_c$

$E_* = \frac{3}{2}(B \cos \beta/e)^{\frac{2}{3}} - z_*$

$z_* = z/y_c$

$y_c = \sqrt[3]{(Q/B)^2/g}$

z = drop in the transition floor at the midpoint (P in Figure 7.7)

y = depth of uniform flow in the approach channel.

When $E_* > \frac{3}{2}$, Equation (7.17) yields two physically real solutions. The solution with the larger magnitude corresponds to subcritical flow conditions in the approach channel. For $E_* \leq \frac{3}{2}$, section 1 acts as the control and the flow depth in the approach channel is critical, i.e.:

$$y_* = 1 \text{ for } E_* \leq \frac{3}{2} \qquad (7.18)$$

Equations (7.17) and (7.18) together with experimental data obtained for a tangential vortex-flow intake are presented in Figure 7.8. The agreement between the predicted and the measured values of flow depth is satisfactory. The details of the experimental study are described in a report by Jain & Kennedy (1983). The dimensions of the tangential intake tested were: B = 0.72 ft (21.9 cm), e = 0.24 ft (7.3 cm), D = 0.96 ft (29.2 cm), z = 0.93 ft (28.5 cm), β = 27.5° and θ = 16.8°. As one would expect, the approach channel depth for the condition $E_* > 3/2$ increases with a decrease in e and z.

7.3.2.2. *Dropshaft diameter*

The dropshaft diameter is governed by the minimum permissible air-core area at the throat for the design discharge. The size of the air core at the throat is determined by treating the throat as another critical, or control, section and using the principle that the velocity of streaming over the critical cross section is equal to that of a small wave in the fluid (Binnie & Hookings, 1948). Let V be the velocity of a 'long' wave of annular form moving in the axial direction on the rotating surface at the throat, and let c be the additional velocity due to wave motion of amplitude η. Then the equation of flow continuity is

$$\pi(a^2 - b^2)\,V = [\pi(a^2-b^2) + 2\pi b\eta]\,(V + c) \tag{7.19}$$

where

a = dropshaft radius

b = air-core radius at the throat.

On the assumption that the tangential velocity at all points across the annulus at the throat is a constant, the increase in pressure head at any radius, due to the spinning portion between radii b and b — η is

$$\Delta h = \frac{v^2}{2}\,[\ln b - \ln (b - \eta)] \tag{7.20}$$

where

v = the tangential component of velocity.

On equating the increase in the pressure head to the decrease in the velocity head, one obtains,

$$\frac{v^2}{8}\,[\ln b - \ln (b-\eta)] = \frac{V^2}{2g} - \frac{(V + c)^2}{2g} \tag{7.21}$$

Eliminating c between Equations (7.19) and (7.21), and taking η as being small gives

$$V = v\left[\frac{(1-\lambda)}{2\lambda}\right]^{\frac{1}{2}} \tag{7.22}$$

where

$\lambda = b^2/a^2$.

The value of V given by Equation (7.22) is the axial-flow velocity at the throat. If it is further assumed that the tangential flow velocity at the throat is equal to the horizontal component of the average velocity at the junction, then

$$v = \left(\frac{gQ}{e}\right)^{\frac{1}{3}} (\cos\beta)^{\frac{4}{3}} \tag{7.23}$$

Eliminating v between Equations (7.22) and (7.23), and using Equation (7.19) yields

$$\sqrt{\frac{(1-\lambda^3)}{2\lambda}} = \left[\frac{Q^2 e}{g\pi^3 a^6 \cos^4\beta}\right]^{\frac{1}{3}} \tag{7.24}$$

Values of λ measured in the laboratory study (Jain & Kennedy, 1983) are compared in Figure 7.9 with those computed from Equation (7.24). The agreement between the observed and computed values of λ is fairly satisfactory, except at low discharges.

A minimum air-core area of 25% of the dropshaft area is commonly adopted to determine dropshaft diameter. The laboratory study showed that the minimum

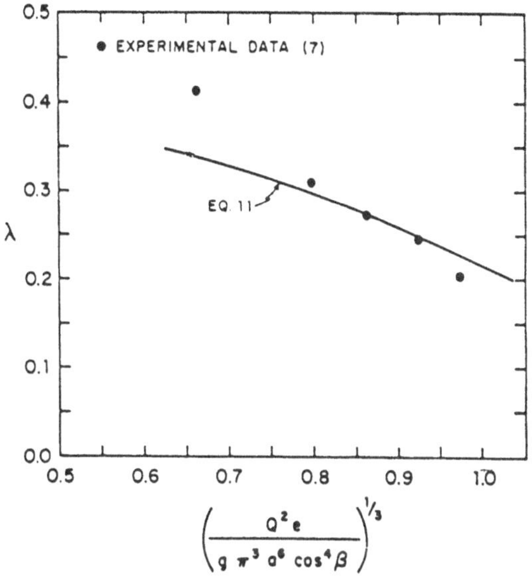

Figure 7.9. Comparison of air-core area.

dropshaft diameter which satisfies the above constraint on air-core area is given by

$$D = \left(\frac{Q^2max}{g}\right)^{\frac{1}{5}} \tag{7.25}$$

Equation (7.25) is based on the experimental data and is in good agreement with Equation (7.24).

The analysis does not account for the effect of sidewall convergence in the approach transition, and includes several tenuous assumptions. Therefore, additional analysis and experimental data are needed to further improve the relationships for designing tangential intakes.

7.4. SUMMARY

Vortex-flow intakes have several hydraulic features which make them very attractive for drop structures that convey water down a vertical or near-vertical dropshaft. The development of the various configurations of vortex-flow intakes is summarized. Hydraulic and relationships for determining the appropriate dimensions of the scroll and tangential forms of vortex-flow intakes are presented.

It is reommended that additional studies on tangential vortex-flow intakes should be conducted to evaluate the effect of various geometrical parameters, such as those describing approach channel geometry, on flow characteristics.

Conclusions 8

G.E.HECKER
Alden Research Laboratory, Holden, Massachusetts, USA

8.1. INTRODUCTORY REMARKS

This volume was prepared with the intent of making hydraulic design engineers conscious of the physical processes involved with swirling and vortex flows, the potential problems which may be encountered (or the benefits to be derived), and methods available to solve anticipated or known problems. As desirable as it may be, there was no intent to provide sufficient information for developing final project designs in lieu of current engineering practice, although reference is given to existing knowledge, guidelines, and examples of past studies such that the design process may be enhanced. This position is taken out of necessity rather than by choice, as an inevitable conclusion is that it is not possible to predict with certainty from basic knowledge, past experience, or guidelines whether or not objectionable swirl and vortices will occur in a given instance. Reasons for this situation are reviewed below.

To provide a means of summarizing the available knowledge represented by this monograph in some organized manner, and to conclude what is important versus unimportant and known versus not known in this field, a series of key questions logically and frequently asked by the hydraulic designer will be posed. The answer to these questions will serve to form the desired conclusions, and this process was believed more fruitful than re-discussing the myriad findings of numerous researchers presented in the monograph.

8.2. SHOULD VORTICES BE PREVENTED?

Although this may seem to be a needless question, it is important to be sure that there will be problems that warrant the studies and probably more expensive structures to prevent the occurrence of vortices. Clearly, there are cases where strong vortices cause objectionable operating problems, including vibration, air binding, a decrease in discharge (i.e. greater entrance losses), lower local pressures (perhaps leading to cavitation), improper machinery performance due to swirl, etc. However, as indicated by Hecker (1981) and Hecker & Larsen (1983), there are also cases where vortices have not caused problems, and it is relevant to remember that a vortex in itself is usually not the issue, but rather the downstream flow characteristics it produces.

For example, if air ingested by a vortex is relatively low in magnitude and stays dispersed as small air bubbles throughout the flow, little consequences may result other than an effective change in liquid density. It is surprising how little air may be ingested by apparently strong vortices. Consider an air core of 1/10 the inlet diameter being pulled with the flow with zero slip, representing a relatively large air core and a conservative assumption regarding air ingestion. From the relative areas involved, it may be surmized that the air ingestion would be only about 1%. Typical values of air ingestion were shown to be less than 2% for 'strong' air core vortices from tests (with $F_h < 0.8$) by Padmanabhan (1984, 1982a). Unless this air is concentrated somewhere in the piping system, little difficulty should result from such low air ingestion. The distance from the inlet to downstream pumps or turbines is a major factor to consider. For installations where prime movers are remote from the inlet, effects of inlet swirl, local low pressure regions, and skewed velocity distributions are dissipated with distance, as quantified by Padmanabhan & Janik (1980). Pipe bends downstream of the inlet may cause as much swirl and unsymmetrical flow patterns as inlet vortices. Increases in water pressure due to elevation changes (such as in penstocks to turbines) decrease the void fraction for a given volumetric air ingestion, and thereby alleviate the resulting downstream effects. Conversely, if the machinery is located close to the inlet, as is the case for the pump intakes discussed in Chapter 6.2, severe detrimental consequences may result from free surface or submerged vortices. This would also be true for typical bulb or tube type hydro turbines which do not have a scroll case to adjust the flow pattern, and where the machinery is directly exposed to poor inlet flow conditions.

There may be other reasons to avoid vortices, from cocerns for small boat safety to asthetic considerations. The main point is to evaluate carefully whether the resulting flow characteristics themselves may prove objectionable for the specific water use under consideration, rather than adopting a universal rule that vortices are to be avoided as a matter of principle.

8.3. WHAT ARE THE PRACTICAL RESULTS OF BASIC RESEARCH?

Chapters 2 and 3 have presented some principles of vortex motion and summarized the results of basic experiments in tanks with bottom outlets. What value is this type of information to the practicing hydraulic engineer?

The main benefit to be derived is an understanding of the relevant flow phenomena, leading to proper evaluation versus blind application of a fixed guide or rule. An example is the appreciation of the important role of circulation in the formation of vortices, and that this factor must be explicitly or implicitly accounted for when considering the submergence required to prevent vortices. This point is directly relevant to the reliability of published guidelines, as will be further discussed in Section 8.5. An understanding of the cause of vortices also leads to more effective means of vortex prevention and dissipation, and understanding the limiting assumptions of analytic techniques leads to proper versus improper application to actual, complex situations.

Appreciating the physics of vortex flow also allows evaluation of (and perhaps even an explanation of) the various seemingly contradictory claims by researchers regarding topics from critical submergence to scale effects. Approach flow characteristics, in terms of detailed velocity distribution and turbulence (implying concern for boundary roughness and approach flow Reynolds number) have direct effects on vortex generation or dissipation, and a lack of knowledge or control of these variables has undoubtedly lead to different experimental results for seemingly similar situations. Experiments using different scales in the same test facility often ignore the influence of changing relative roughness and turbulence, the latter being affected not only by a change in Reynolds number, but also by relatively different turbulence scales produced by fixed flow distributors. Approach flow turbulence may also produce the 'unexpected' trend of less vortex activity at higher withdrawal flow rates.

Without a careful delination of vortex strength versus time, it is not possible to make a proper comparison of one situation to another, be it a case of defining a critical condition versus a non-critical condition, or comparing the results of experiments at two different scales, including model to prototype. Since vortices are frequently not steady with time (particularly at so-called 'critical' conditions), a statistical approach such as shown in Figure 2.3 is frequently necessary to properly compare data, and this requires a vortex classification system such that the vortex strength versus time can be recorded. A statistically meaningful level of vortex activity, such as the time average strength or that strength exceeded only 10% of the time, can then be selected when plotting vortex related parameters. Without this effort, misleading trends may appear.

A basic understanding also helps in evaluating data on the 'critical' submergence, that submergence when an air core to the inlet first forms. In a typical experimental facility a change in water level does not produce a change in flow pattern and higher submergences usually result in less vortex intensity at a given flow. In many actual (prototype) installations, this is not the case. Not only does a change in flow pattern and therefore circulation frequently occur with changes in water level, but higher water levels frequently provide dead water regions where swirl can build with time. For example, it is not unusual for some pumped storage projects to have as much (or more) vortex activity at high water levels as at lower submergences, as shown by Hecker (1984). Appreciating the influence of parameters controlling vortex motion, as learned from basic experiments, helps in explaining these seemingly contradictory findings.

8.4. ARE MODELING SCALE EFFECTS UNDERSTOOD?

As hydraulic model studies will continue to be used to predict the occurrence and intensity of vortices, it is proper to question the reliability of models as predictors of prototype performance. Assuming there are no errors in model design, construction, or operation, this question primarily concerns possible scale effects in modeling vortices and related swirling flow phenomena, both surface and submerged.

Chapters 3 and 5 have summarized the results of various investigations on scale effects. But what is the essence of this information, what is critical, and what is of negligible impact? It is clear from the dimensionless equations of motion that a geometrically similar model will scale vortices if the Froude number, circulation number, Reynolds number, and Weber number are all equal in model and prototype. As with any free surface flow model, Froude numbers between model and prototype must be equal since the governing forces of inertia and gravity are thereby scaled. Flow patterns, and therefore the circulation number, will thereby be equal if the Reynolds and Weber number are sufficiently high. If they are insufficient in magnitude, scale effects result. But how large a magnitude is sufficient? Before addressing this point, a factor previously mentioned should be explored further. The importance of the circulation number is obvious, as circulation causes the vortex. Since the resultant circulation at an outlet depends on the approach flow pattern, anything contributing to scaling the approach flow pattern must be carefully examined. This includes not only complete geometric similitude, but also the relative roughness of the boundary and the approach flow Reynolds number. Proper attention must be given to the friction factor and turbulence (including waves) of the approach flow, and insufficient concern has been placed on scaling the approach flow as compared to scaling the vortex itself.

Assuming approach flow characteristics are scaled, determining the critical values of the Reynolds and Weber numbers can be settled, at least for engineering purposes. There are numerous studies which have addressed this point, but the answer is not quite so obvious as may first appear. As the Froude, Reynolds, and Weber number all contain the reference velocity (or flow) it is difficult to separate out the effects of one variable since a meaningful change in Reynolds number and Weber number cannot be made by changing only the fluid viscosity or surface tension, respectively. A plot of a vortex controlling variable, such as the circulation number for an incipient air core vortex, versus either the Reynolds or Weber number does not indicate a unique dependence due to the influence of the Froude number. However, if the vortex characteristic does not change on such a plot with either the Reynolds or Weber number, then at least an upper bound for the critical or limiting values of these numbers may be available. This has been done by various people, such as by Anwar et al. (1978), who indicate a minimum value for the model radial Reynolds number, $R_R = Q/\nu h$, of about $3{\times}10^4$, and a minimum value for the square of the model submergence Weber number, $W_h^2 = \rho h V^2/\sigma$, of about 10^4.

Support for the stated value of the Reynolds number comes from experiments by others, such as Daggett & Keulegan (1974), Padmanabhan & Hecker (1984), and Zielinski & Villemonte (1968), and from model to prototype comparisons, such as by Dhillon (1979) and Hecker (1981). There is some evidence that lower values of the Weber number may be adequate for a model study as shown by the experiments of Daggett & Keulegan (1974), Jain et al. (1978), and Padmanabhan & Hecker (1984).

Critical values of a Reynolds number for submerged vortices may be easier to derive since the Froude number is eliminated as a variable. However a new difficulty arises, that of quantifying the strength of submerged vortices. In a series of experiments on various scale models of depressed floor sumps, Padmanabhan & Hecker (1984) reported that submerged vortex intensity was approximately proportional to the rotation rate of a swirl meter in the outlet pipe since free surface vortices were negligible for those test conditions. The submerged vortices were caused by flow separation at the edge of the sump depression, and it is believed that the resulting submerged vortices were dependent on an approach Reynolds number, based on the approach flow velocity and depth. Figure 2.7 by Padmanabhan & Hecker (1984), shows a clear dependency of the swirl intensity on this Reynolds number, and indicates that the approach flow Reynolds number should be above approximately 3×10^4 to avoid potential scale effects due to separation induced swirl.

As vortex characteristics become asymptotically independent of scale effects it is not of practical importance to debate the exact values of the Reynolds and Weber number limits for free and submerged vortices. In addition, the investigator has the freedom to choose what vortex condition may be acceptable, and some variable degree of conservation may be applied depending on the model values of the Reynolds and Weber numbers. Considering also that model studies are used to eliminate strong vortices from the final design, available quantification of scale effects for air-core vortices seems sufficient.

It is reasonable that less scale effects exist for weaker vortices and this seems to be the case based on the model prototype comparison by Hecker (1981). However, since various means of overcoming and evaluating potential scale effects have been proposed, it is worth synthesizing these methods into a general framework.

One traditional way of evaluating scale effects and predicting prototype behavior is to contruct a number of models at different scale ratios. Models at larger scale ratios yield higher Reynolds and Weber numbers at the same Froude number. Given Froude scale test conditions corresponding to three models may be represented by points A, B, and C on Figure 8.1. A more frequent and less costly technique is to increase the flow in a model while maintaining the scaled submergence and outlet size, thereby inreasing the model velocities proportionally. Such increased velocity test conditions may be represented by points a_1, through a_3 on Figure 8.1, where a test point, a_4, is also shown to represent the condition of setting the prototype intake velocity in the model. Chemical additives to decrease the water viscosity, or increases in model water temperature to achieve the same effect

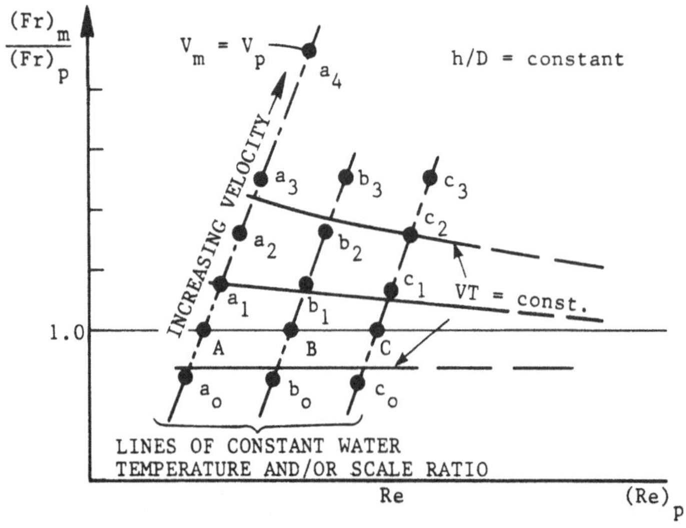

Figure 8.1. General evaluation of scale effects.

(while also reducing the surface tension), result in moving from test points A to B or C in a given Froude scaled model, or from a_1 to b_1 or c_1, etc., at increased flows. For completeness in evaluating scale effects, the model may be operated at less than Froude scaled flows, as indicated by points a_o, b_o, and c_o. For any model operating condition discussed above, and reached by any combination of model scale ratio, velocity ratio, water temperature, or chemical additives, a characteristic feature vortex may be noted, such as the vortex type occurring a certain percent of time or the critical submergence for an incipient air core vortex. Lines of constant vortex characteristic may then be superimposed, as shown on Figure 8.1 using a constant vortex type. If such lines are horizontal, i.e. independent of Reynolds number, no viscous scale effects exist and any observed changes in vortex intensity are due to Froude number changes. If there is some dependency of vortex type (or other selected parameter) with Reynolds number at Froude scaled conditions, scale effects exist and a projection to prototype operating conditions could be made.

It may be concluded from Figure 8.1 that most of the commonly used techniques to overcome potential scale effects may be represented by test conditions on an Fr — Re plane for a constant relative submergence. Any decreases of the inlet size to increase inlet velocities would change the relative submergence, a primary geometric similitude parameter. It can be misleading to operate at only one preselected test condition off the (Fr) model = (Fr) prototype line since the different vortex condition observed may not be indicative of prototype conditions due to the different Froude number in the model. Also, the approach flow pattern and associated circulation may be altered since the Froude similitude criterion is not satisfied, clearly a primary requirement of open channel flow models. Although operating a model at somewhat higher velocity than given by Froude scaled flow may yield conservative results if increased flows produce increased vortex intensity,

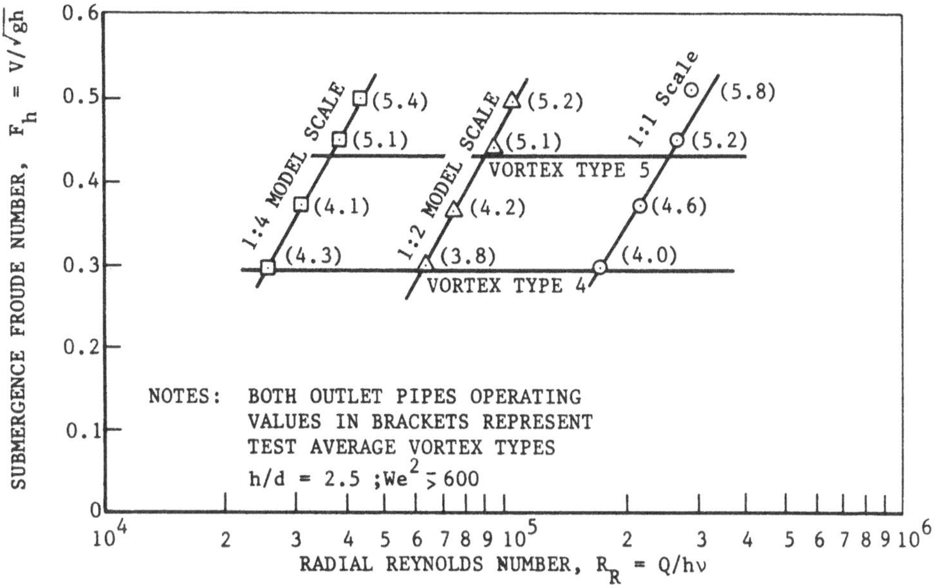

Figure 8.2. Test average vortex strength from various model scales of a depressed sump (constant flow pattern) (Padmanabhan & Hecker, 1984).

this is an arbitrary adjustment factor varying in magnitude with the situation under study. Just as viable a safety factor is the investigators' choice of what vortex type will be allowed for what percent of time. The concept of operating models at prototype velocity appears to produce erroneous results in most cases as discussed by Dhillon (1979) and Hecker (1981), and has little basis for models operated above the required limiting Reynolds and Weber numbers.

A recent study of vortex scaling in a depressed sump by Padmanbhan & Hecker (1984) involved a family of models at different scale ratios, each operated over a range of flows. Statistical evaluation of the average vortex strength at each flow rate, using the classification scale given in Figure 2.2, resulted in the data shown in Figure 8.2. It may be seen that there was no increase in vortex strength with Reynolds number for a given Froude number, and that at a fixed Reynolds number, the vortex strength only increased due to the increase in Froude number. A plot of average vortex strength versus only the Reynolds number would mask the influence of the Froude number. Figure 8.2 supports the previously mentioned minimum value for the radial Reynolds number of 3×10^4.

The ultimate test of scale effects is to compare the vortex characteristics of models to the prototype operating under comparable conditions. This has recently been done by Hecker (1981) for a variety of intakes for which sufficient model and prototype data are available. In general, it was found that models operated at Froude scaled flows scaled weak vortices well, and no case was found where weak model vortices translated into strong prototype vortices. Based on the results of a few model prototype comparisons, and also considering the results of some basic experiments, it was concluded that some scale effects were evident for strong air-core vortices, but this conclusion was tempered by insufficient data for the projects in question for a detailed comparison. One of these comparisons was, however, based on detailed observation of both the model and the prototype, and this case was recently elaborated on by Hecker (1984). It may be summarized that the reported scale effects occurred mainly when the model radial Reynolds number was somewhat below the previously stated limiting value, and that relatively small increase in flow, a factor of about 2 for the 1:50 scale model, was sufficient to overcome the scale effects regarding vortex persistence.

In summary, there seems little doubt that a model may be designed and operated to have negligible scale effects for engineering purposes, even for air-core vortices, if care is taken in simulating the approach flow and if the model has sufficiently high Reynolds and Weber numbers.

8.5. IS THE INTAKE GEOMETRY IMPORTANT?

The intake geometry includes factors such as the distance to adjacent boundaries (wing walls, floors, roofs, and embankments), the direction of the outflow, the type of piers and racks used for gates and trash screening, and details of the entrance which could cause flow separation versus rounded entrances with gradual area reductions. The aspect ratio of the intake area may also be relevant for high aspect ratios. For example, an intake slot (circular or straight in plan) may perform differently (at the same relative submergence) than a square or circular opening of the same area. These features have effects on the local flow pattern which may influence the generation or dissipation of swirl and associated free or submerged vortices. Examples of actual intakes are given in the figures of Chapter 6.1, and by Pennino & Hecker (1979). The range of intake geometries (in addition to the effects of approach flow conditions) is one reason for the varying guidelines for the critical submergence.

Most basic studies have been conducted with square edged circular intakes, although Anwar et al. (1978) conducted some tests with bell-mouth intakes with little effect. The effect of the wall clearance on the critical submergence for vertical upward withdrawal has been tested by Dicmas (1967), Denny (1956), and Anwar & Amphlett (1980). Those tests which were related to circulation are summarized on Figure 8.3, clearly showing that as the back or side wall clearance is reduced, the critical submergence (based on free surface vortices) is also reduced, and that this reduction in critical submergence is rather dramatic for cases with moderate and strong circulation. Presumably, the proximity of a wall(s) suppresses free surface vortices, allowing a lower water level without vortex formation. However, Dicmas (1967) has shown a weak reverse trend between critical submergence and sidewall clearance when no other geometric parameter was changed. Both Denny (1956) and Dicmas (1967) showed that the critical submergence decreases as more clearance is provided between a vertically upward intake and the floor. Since it is also known that the proximity of boundaries affects energy losses and submerged vortices,

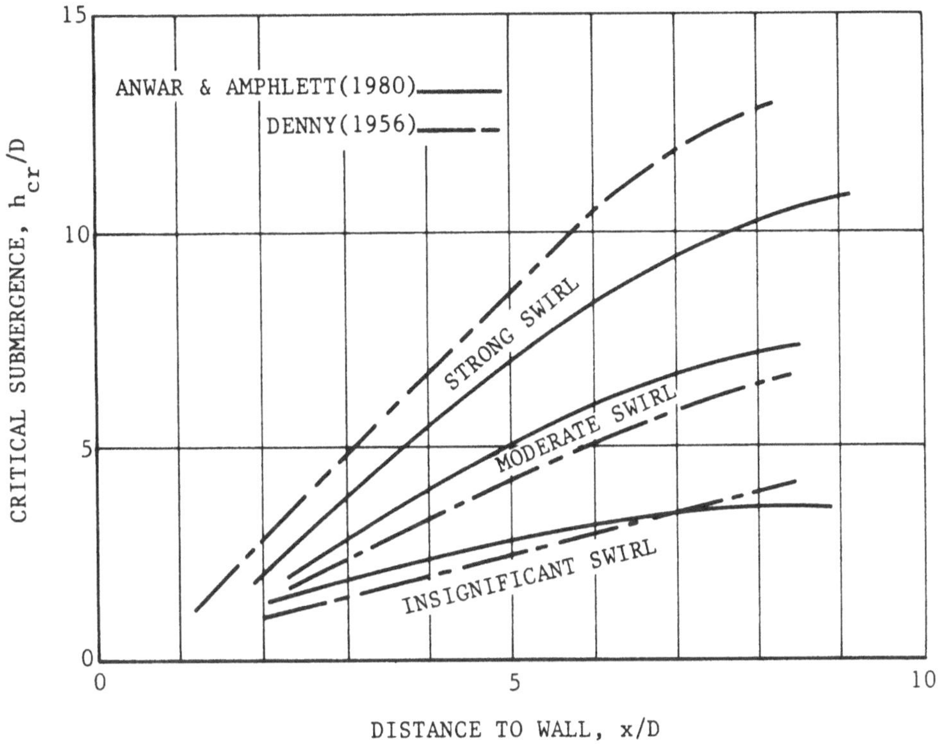

Figure 8.3. Effect of wall clearance on the critical submergence for vertically upward intakes.

it seems evident that boundary clearance is an important geometric variable.

The question of intake withdrawal direction on critical submergence has been thoroughly investigated by Knauss in Chapter 4. It seems evident that the withdrawal direction has some influence on the critical submergence, although this is but one geometric factor describing an intake. Due to the varying influence of other important experimental variables, such as the distance to adjacent boundaries and the subjective method of defining the critical submergence based on observing time varying vortex strength, it may be appreciated that the most reliable evaluation of the withdrawal direction effect is obtained if the same facility is used by the same personnel. Conversely, it is not as reliable to use data from various sources to evaluate the effect of withdrawal direction.

Tests on horizontal versus vertical intakes have been reported by Anwar & Amphlett (1980), and these are compared in Figure 4.11 in detecting a 20% increase in critical submergence for a horizontal withdrawal compared to a vertically upward withdrawal. Unfortunately, the data on which this difference is based are mainly in the region where the radial Reynolds number is below the stated limiting value of 3×10^4, and the difference at higher Reynolds numbers is less. The effect of a horizontal withdrawal versus a vertically downward direction has been tested by Padmanabhan (1982a, b), who compared the relationship between the average

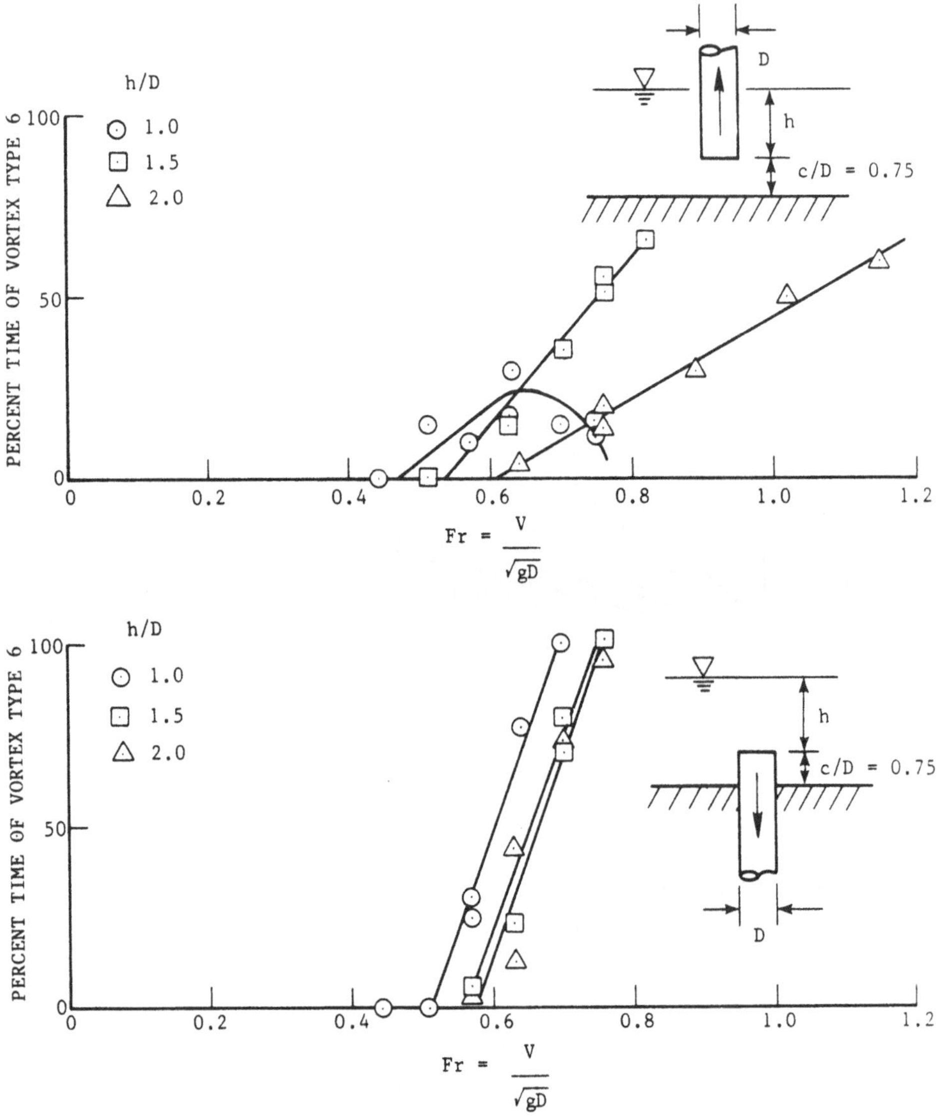

Figure 8.4. Frequency of air core vortex in vertically upward versus vertically downward intake (same facility — $R_R = 3 \times 10^4$ to 1×10^5).

vortex type and the submergence for two different sump sizes with the same approach flow pattern. For the shallower sump, similar strength vortices were produced by the vertically downward versus the horizontal withdrawal, while for the deeper sump, there was a scattered tendency for slightly stronger vortices with the vertically downward withdrawal.

Additional tests were recently conducted at the Alden Research Laboratory of WPI to investigate the difference between a vertically upward versus a vertically downward withdrawal in an otherwise identical experimental facility. Figure 8.4

shows a comparison between the frequency of a Type 6 vortex (see Figure 2.2) versus intake Froude number for the 180° change in withdrawal direction, and some interesting observations may be made. For the lowest tested submergence with the vertically upward withdrawal, the persistence of air-core vortices did not continually increase with increased Froude numbers. A trend of decreasing vortices with increasing Froude numbers has been reported previously by Hecker (1984); this phenomena was discussed in Section 8.2. The manner of how the critical submergence in defined is shown to be important in that similar critical submergences would result at a given Froude number, if an incipient Type 6 (i.e. near zero frequency) were used, whereas a higher frequency criterion (such as 50%) would result in reporting a higher critical submergence for the vertically downward suction. The Froude number at which Type 6 vortices start is similar in both withdrawal cases, but the increase in frequency of vortexing with increasing Froude number is less for the vertically upward case. These last two factors indicate the inherent tendency for air-core vortices to form is similar, but that the vertically upward geometry attenuates the persistent formation of air core vortices. This is reasonable considering the withdrawal column is a physical obstruction displacing the vortex and forming an adjacent boundary with probable related energy dissipation. Also, irregular vortices are shed from the suction column, causing turbulence and disrupting the formation of a stable vortex, and the length of the air core is slightly increased, all compared to a vertically downward suction.

It is obvious that there are many other geometric variables besides the proximity to boundaries and the direction of withdrawal elaborated on above, but little quantitative information is available regarding these additional geometric variables. Since the effects of geometric variables on the flow interact, it may be difficult to isolate one parameter in a general manner since other geometric parameters may mask the effects in certain ranges. For example, splitter walls may be used to dissipate intake swirl and vortices, and the presence of intake piers may therefore control the flow pattern in spite of other undesirable geometric or flow features. It may be stated, in summary, that the inlet geometry is an important variable, and is one reason why the reliable prediction of vortices by guidelines continues to be difficult.

8.6. DO PUBLISHED GUIDELINES WORK?

Various investigators have proposed guidelines for designing vortex-free intakes. Essentially all guidelines give the ('critical') submergence required to avoid air-core vortices as a function of some independent intake variable. The more practical of these guidelines give the required submergence in terms of the intake diameter, Prosser (1977), flow, as in the Hydraulic Institute Standards (1975), the intake velocity, Denny & Young (1957) and Gordon (1970), or an intake Froude number, Peterson & Noble (1982), Quick (1970), and Reddy & Pickford (1972), all without explicitly including a measure of circulation. The possible exception is the guideline of Gordon (1970) which considers circulation by giving two trends of required submergence, one for 'symmetrical' approach flow and one for intakes with a 'lateral' approach flow. More basic guidelines, such as by Anwar et al. (1978), see Figure 2.4, and Jain et al. (1978), require explicit knowledge of the circulation, a factor usually not known a priori. Also, as was discussed in Section 8.4, the intake geometry has an important but poorly defined influence. Available guidelines are thus difficult to apply with confidence since it is not clear if they are applicable to the geometry under consideration, and because the circulation is either not sufficiently accounted for or must be known.

Guidelines should be subjected to an independent test of reliability using other intakes than included for origination of the guidelines. Dhillon (1979) in a comprehensive evaluation of various modeling techniques and comparable prototype performance indicates a poor correlation with the available guidelines. Pennino & Hecker (1979) have evaluated model studies of a number of pumped storage upper reservoir intakes for which the original vortex prone designs and the final non-vortexing designs were available, and their summary showed little dependence on submergence. Possible reasons for vortices at high submergences is that circulation may also be high under these conditions due to the influence of approach topography, and that larger stagnant areas exist where vortices can build with time. Their data suggested that a possible alternate guideline, for cases where the circulation is unknown but within practical limits, is for the submergence Froude number (based on submergence rather than the intake diameter) to be less than about 0.25. This is in agreement with Figure 2.5, which shows that a submergence Froude number of less than about 0.25 eliminates air ingestion due to surface vortices even for conditions where strong local circulation was generated by forcing an unsymmetrical approach flow. This does not indicate a lack of swirl and vortices, but does ensure that any surface vortices will not lead to measurable air ingestion. The safety margin inherent in this simple guide can be appreciated by the transformation of

$$F_h = V/\sqrt{g\,h} \leqq 0.25 \tag{8.1}$$

to

$$h/D \geqq 16\ V^2/gD = 16\ Fr^2 \tag{8.2}$$

If an intake can economically be constructed to this guide, it may be concluded that no air drawing vortices will occur. Such a design may, however, be prohibitively expensive, especially for large hydraulic structures.

To provide an independent test of those published guidelines which are based on considering a variety of intakes, the previously mentioned pumped-storage intakes evaluated by Pennino & Hecker (1979) are compared to the guidelines on Figure 8.5. A listing of the included projects is given in Table 8.1. It may be seen that final designs with (or without) air core vortices may have low or high submergences due to the unspecified influence of intake geometry, approach flow circulation,

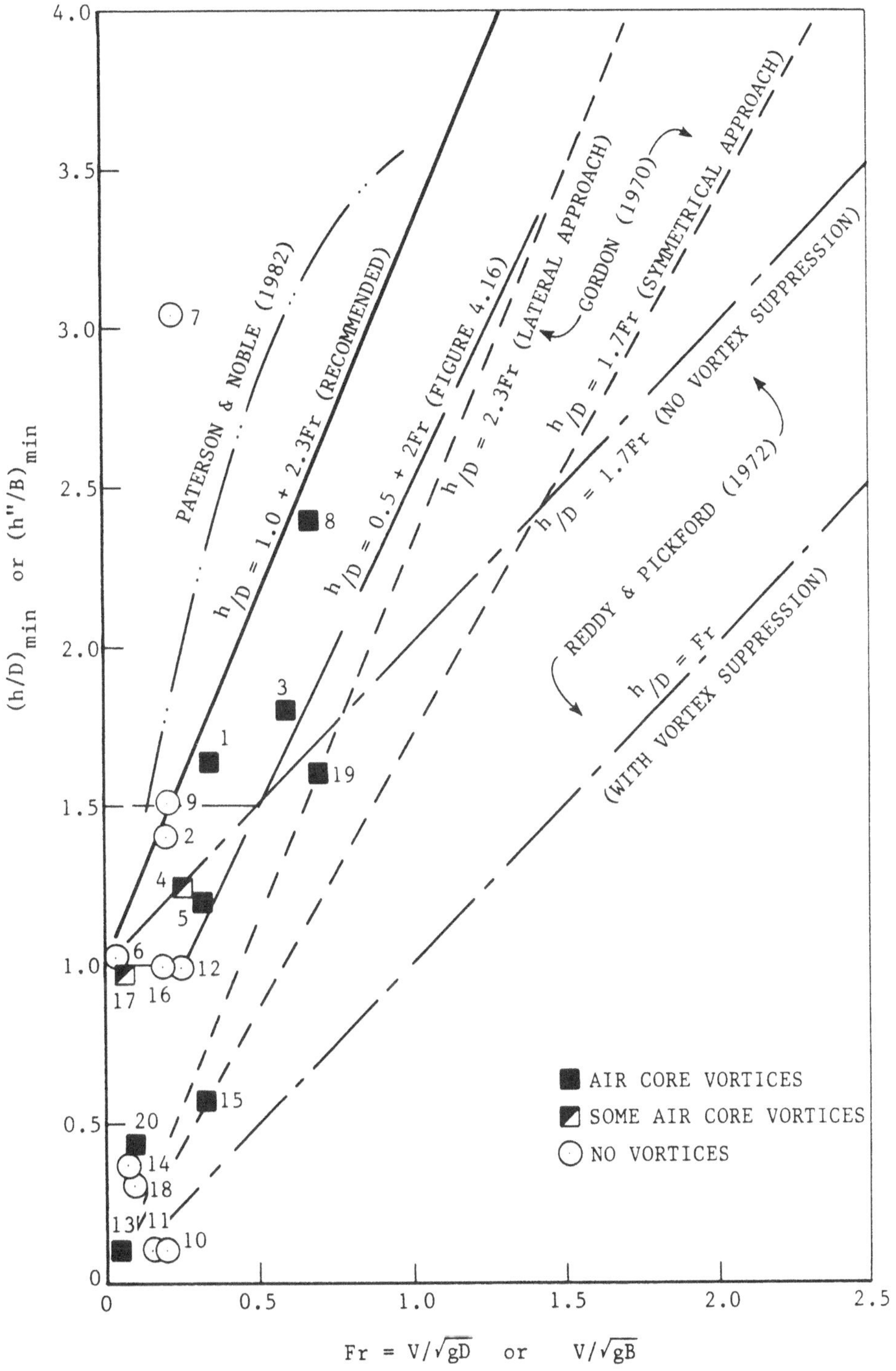

Figure 8.5. Intakes with and without objectionable vortices compared to available guidelines.

Table 8.1. Pumped-storage intake structures with and without vortices (Pennino & Hecker, 1979).

Project name	Air core Yes	Air core No	V (ft/s)	D or B[a] (ft)	h_{min}[b] (ft)	Fr[c]	h_{min}/D	Intake type and comments
1. Bad Creek (original)	x		7	13	21	0.35	1.6	Covered – vertically down (vortices up to h/D = 3.7)
2. Bad Creek (final)		x	8	51	70	0.20	1.4	Open – vertically down (with anti-vortex piers)
3. Bear Swamp (original)	x		18	28	50	0.60	1.8	Open – vertically down (simple shaft)
4. Bear Swamp (final)	Some		9	40	50	0.29	1.2	Open – vertically down (borderline vortices)
5. Blenheim Gilboa (original)	x		12	38	45	0.33	1.2	Open – vertically down
6. Blenheim Gilboa (final)		x	1.2	28	27	0.04	1.0	Covered – vertically down
7. Cabin Creek (final)		x	4	9	29	0.22	3.0	Covered – vertically down
8. Cornwall (original)	x		24	40	95	0.67	2.4	Open – vertically down (simple shaft)
9. Cornwall (final-model)		x	10	62	91	0.22	1.5	Open – vertically down, with piers and trash rack
10. Davis (final)		x	7.6	47	5	0.20	0.1	Horizontal; with piers
11. Fairfield (final)		x	7.6	53	6	0.18	0.1	Horizontal; with piers
12. Jocassee (final)		x	6.3	20	20	0.25	1.0	Horizontal – openings around cylinder; with piers
13. Kinzua (original)	x		1	120	15	0.02	0.1	Open – vertically down (strong vortices to h/D = 0.5)
14. Kinzua (final)		x	1.7	19	7	0.07	0.4	Covered – vertically down
15. Ludington (original)	x		11	35	20	0.33	0.6	Horizontal – six; in line (final design uses large perforated wall to suppress vortices)
16. Muddy Run (final)		x	4	15	15	0.18	1.0	Horizontal – multiple openings around cylinder; with piers
17. Northfield (original)	Some		2.2	21	20	0.08	0.9	Covered – downward at angle (borderline vortices)
18. Northfield (final)		x	5	80	25	0.10	0.3	Horizontal – single opening
19. Raccoon Mt. (original)	x		23	35	55	0.69	1.6	Open – vertically down (final design uses large perforated tower to suppress vortices)
20. Taum Sauk (final)	x		4	46	20	0.10	0.4	Open – vertically down

(a) For horizontal intakes, B is height of intake, irrespective of geometry.
(b) h_{min} is either h or h″ (Figure 4.1) at minimum water level.
(c) Froude number with D if circular or B if horizontal intake.

and local vortex suppression due to piers, trash racks, etc. Correct prediction of these intakes regarding vortex formation would not have been achieved by the guidelines, with the exception of the submergence(s) recommended by Paterson & Noble (1982) for pump intakes. A similar conclusion was reached by Gulliver et al. (1983), who used the data of Gordon (1970), Pennino & Hecker (1979), and additional projects to determine that the previously published guidelines would not guarantee negligible vortices and that the submergence per se could not be used to distinguish between acceptable and unacceptable intakes. The newly suggested guide by Knauss shown on Figure 4.16 is also seen to not predict sufficient submergences required by some of the listed projects nor by some of the projects reviewed by Gulliver et al. (1983).

In formulating a more conservative guideline to account for the data on Figure 8.5, a number of factors were considered. A combination of the envelope approach used in Figure 2.5, the linear dependency of critical submergence with Froude number suggested by Knauss in Chapter 4, and the slope of the 'lateral approach' guide suggested by Gordon (1970) leads to:

$$(h/D)_{cr} \text{ or } (h''/B)_{cr} = 1.0 + 2.3\ Fr \qquad (8.3)$$

where

h = submergence above a vertical intake

h'' = submergence above the roof of a horizontal intake of height B (see Figure 4.1).

This recommendation, shown on Figure 8.5, does not provide for the submergence suggested by Paterson & Noble (1982) for pump intakes, or for the linearized version of that curve suggested by Padmanabhan in Chapter 6.2, due to the implied expense for large hydraulic structures, especially horizontal intakes. For pump intakes however, it might be practical to increase the submergence by 0.5D without incurring much extra expense. It should also be noted that Prosser (1977) suggests a minimum (fixed) submergence of 1.5D for pump intakes. As additional submergence can only help in minimizing vortices (all other factors being equal), the designer may increase the submergence above the guideline if resulting project costs remain reasonable.

Such 'envelope' type limits as Equation (8.3) should be refined with time as more data become available. None of the above suggestions nor the previously published guidelines explicitly account for circulation at a given site, except that reasonable circulation magnitudes are included in the data on which these suggestions and guidelines are based. It must, therefore, be concluded that even Equation (8.3), which is more conservative than previously published guidelines, should not be used to select a required submergence when it is critical that no strong vortices occur. Although the lack of reliable guidelines is unfortunate, it would be more so to completely rely on them. If the prediction of surface vortex strength is critical, and if submerged vortices may cause serious problems, a hydraulic model study is necessary. This is also true when considering a conclusion reached in Section 2.1 that analytical techniques can not yet be relied on for vortex predictions.

8.7. HOW MAY VORTICES BE DISSIPATED?

During a model study to produce a vortex-free design, or in cases where objectionable vortices occur in the prototype, various techniques may be used to dissipate vortices to negligible strengths. The techniques vary depending on whether free or submerged vortices are a problem, and whether a model or prototype intake is being addressed. Some changes to an actual intake are clearly more difficult or even impractical compared to developing a design through use of a model. In all cases, one basic concept involves reducing the circulation at the intake, either through adjusting the approach flow or through local energy dissipation. In addition, reductions in inlet velocity are beneficial. A factor often overlooked is that a poor velocity distribution at the intake face produces an effective higher local velocity and susceptibility to vortexing. The inlet velocity should be as uniform as possible.

In general, a preferred intake geometry will minimize flow separation, swirl and circulation, and provides for a uniform approach flow which continuously accelerates toward the intake opening. No regions of stagnant water should exist near the intake, as these will provide locations where circulation may build with time.

For multiple intakes in which one or more openings may have no flow while others are operating, unfavorable swirling flow patterns may be produced under such partial operation conditions, even if the intake operates well under full flow. This factor must be explored if potential problems are to be avoided.

8.7.1. *Free surface vortices*

The physical relationship of the intake relative to the approach flow may be changed early in the design stage to improve the uniformity of the approach flow, thereby minimizing velocity variation and circulation. Guide walls may be altered to provide for a uniform accelerating flow and designed so as to fill in any regions of stagnant water susceptible to rotation. If vortices persist, the surface flow pattern must be altered, and one effective method employs a vertical 'curtain' wall projecting downward into the water some distance upstream of the problem area. Another way to dissipate surface rotation is by use of a horizontally oriented grid at or slightly below the surface. Changes at the intake opening are also effective in some cases, and the use of grids, solid horizontal projections or 'hoods', and splitter vanes should be considered. If the inlet(s) may be lowered in elevation at reasonable cost, the greater submergence may also be beneficial, assuming no increase in circulation results.

8.7.2. *Submerged vortices*

Submerged intake vortices must end at a solid boundary or be connected to another adjacent intake. The strength of such vortices may be reduced by increasing the distance between the intake and boundary or between intakes. Increasing the roughness of the boundary may also be effective, as rotational energy is thereby dissipated. Gratings around the opening, or vanes within the opening are also effective in dissipating rotating flow patterns. All other considerations regarding flow uniformity and the undesirability of stagnant flow areas discussed above also apply for submerged vortices.

8.8. RECOMMENDATIONS FOR FUTURE RESEARCH

It is encouraging that considerable knowledge exists on the basic characteristics of vortices, as this is of value to the practicing hydraulic engineer in avoiding problems with unusual designs. Even though neither mathematical models nor available guidelines can reliably predict the occurrence or non-occurrence of vortices, enough is known that proper decisions can be made regarding the need for site specific studies. It seems evident that enough is also known about possible scale effects such that a proper hydraulic model can be designed and operated to reliably predict field conditions.

Rather than continuing experiments to define the basic features of surface vortices, and to quantify scale effects of air-core vortices (usually eliminated by hydraulic model studies), it is recommended that research on the analytical prediction of circulation be conducted. If the overall flow pattern around an intake can be defined, appropriate boundary conditions to analytical models would be available, and reasonable predictions of vortex intensity would be possible. However, due to the influence of the local intake geometry, additional work is also necessary on basic geometric effects.

The other area requiring considerably more research concerns submerged vortices. For some installations, these are more detrimental than surface vortices, yet little is known about predicting the occurrence of submerged vortices, their intensity, and scale effects.

References

Ables, J.H. 1969. *Vortex Problem at Intake, Lower St.Anthony Falls Lock and Dam, Mississippi River, Minneapolis, Minnesota; Hydraulic Model Investigation.* AEWES, Vicksburg, MS., Tech. report. no. HL-79-9.

Ackers, P. & E.S.Crump 1960. The Vortex Drop. *Institution of Civil Engineers* 16(4).

Allen, J. & G.A.Taylor 1958. A Note on Free Vortices Observed in Scale Model Tests of a Bellmouth Overflow Spillway. *Proc. of the Institute of Civil Engng., London, UK*, vol. 10.

Amphlett, M.B. 1976. *Air-entraining Vortices at a Horizontal Intake.* HRS Wallingford, Report no. OD/7.

Amphlett, M.B. 1978. *Air-entraining Vortices at a Vertically Inverted Intake.* HRS Wallingford, Report no. OD/17.

Amphlett, M.B. 1979. Vortex Formation at Vertical Pipe Intakes; Discussion of Jain et al. (1978). ASCE, *J. Hydraul. Div.* HY 10.

Anderson, S.H. 1961. *Model Studies of Storm-Sewer Drop Shafts.* St.Anthony Falls Hydraulic Laboratory, Univ. Minnesota, Minneapolis, Minn., Technical paper no. 35, Series B.

Anwar, H.O. 1965. Flow in a Free Vortex. *Water Power* 1965(4).

Anwar, H.O. 1966. Formation of a Weak Vortex. *Journal IAHR* 1966(1).

Anwar, H.O. 1967. Vortices at Low Head Intakes. *Water Power* 1967(11).

Anwar, H.O. 1968a. Vortices in a Viscous Fluid. *Journal IAHR* 1968(1).

Anwar, H.O. 1968b. Prevention of Vortices at Intakes. *Water Power* 1968(10).

Anwar, H.O. 1969. Turbulent Flow in a Vortex. *Journal IAHR* 1969(1).

Anwar, H.O. 1983. The non-dimensional parameters of free-surface vortices measured for horizontal and vertically inverted intakes. *La Houille Blanche* 1983(1).

Anwar, H.O. & M.B.Amphlett 1980. Vortices at Vertically Inverted Intake. *Journal IAHR* 1980(2).

Anwar, H.O., J.A.Weller & M.B.Amphlett 1978. Similarity of Free-Vortex at Horizontal Intake. *Journal IAHR* 1978(2).

Babb, A.F. & W.C.Mih 1984. Selection of Model Test Programs with Reference to Viscous and Gravity Force Similitude. *Proceedings of the IAHR/DVWK Symposium on Scale Effects in Modelling Hydraulic Structures*, Paper 6.2. Technische Akademie Esslingen.

Baker, D.W. & C.L.Sayre 1974. Decay of Swirling Turbulent Flow at Incompressible Fluids in Long Pipes. In *Flow – Its Measurement and Control in Science and Industry*, vol.2, pp. 301-312. Instrument Society of America.

Berge, J.P. 1966. A Study of Vortex Formation and Other Abnormal Flow in a Tank with and without a Free Surface. *La Houille Blanche* 1966(1).

Bhargava, P.K. 1982. Vortex Formation at Pipe Offtake from an Open Channel. M.E. thesis, University of Roorkee, Roorkee.

Bhargava, P.K., N.Vittal & K.G.Ranga Raju 1984. Vortex Formation at Pipe-Offtake in an Open

Channel, in Channels and Channel Control Structures. *Proceedings of the 1st International Conference on Hydraulic Design in Water Resources Engineering.* Springer, Berlin.

Bhatt, V.P. & B.P.Shah 1981. Problems of Simulation of Vortices in Models. *IAHR Congress New Delhi*, Subject D(a), Paper no.12.

Binnie, A.M. & D.P.Harris 1950. The Application of Boundary-Layer Theory to Swirling Liquid Flow through a Nozzle. *Q. J. Mech. Appl. Math.* 3(1):89-106.

Binnie, A.M. & G.A.Hookings 1948. Laboratory Experiments on Whirlpools. *Proc. Royal Society of London*, series A, vol.194.

Bisaz, E., K.-C.Taubmann & P.Fischer 1976. Improvement in Transport of Floating Debris. *Water Power* 1976(11).

Blaisdell, F.W. 1960. Hood Inlet for Closed Conduit Spillways. ASCE. *J. Hydraul. Div.* HY 5.

Blaisdell, F.W. 1979. Vortex Formation at Vertical Pipe Intakes; Discussion of Jain et al. (1978). ASCE. *J. Hydraul. Div.* HY 10.

Boillat, J.L., J.Bruschin & D.Rolewicz 1984. A Boundary Layer Suction Vortex Suppressor. *Proceedings of the IAHR/DVWK Symposium on Scale Effects in Modelling Hydraulic Structures,* Paper 6.6. Technische Akademie Esslingen.

Boughton, N.O. 1959. Laboratory Studies on the Use of Siphons and Vortex Chambers for the Exclusion of Air at Secondary Intakes. *IAHR Congress Montreal*, vol. 2.

Bozoian, P.M. & J.Larsen 1977. *Model Studies of Intake Pumphouse for the Perry Nuclear Power Plant Cooling Towers.* Alden Research Laboratory, Worcester Polytechnic Institute, Holden, Mass., Report no.153-177.

Bretschneider, H. 1981. Massnahmen zum Vermeiden des Lufteintrags bei Einlaufbauwerken. *Wasserwirtschaft* 1981(5).

Brocard, D.N., C.H.Beauchamp & G.E.Hecker 1982. *Analytic Predictions of Circulation and Vortices at Intakes.* Electric Power Research Institute, Research project 1199-8, Final report.

Brooks, N.H. & W.H.Blackmer 1962. *Vortex Energy Dissipator for San Diego Ocean Outfall-Laboratory Investigations.* W.M.Keck Laboratory of Hydraulics and Water Resources, California Institute of Technology, Pasadena, California, Report no.KH-R-5.

Bruschin, J. 1985. Modelling vibration and vortex formation at Piedra del Aguila. *Water Power* 1985(5).

Bucher, K.G. & J.Amorocho 1968. *Hydraulic Model Study of the Shaft Spillway for Del Valle Dam.* University of California, Davis, Dept. Water Science and Engineering, Water Science and Engineering Papers no.1020.

Burgi, P.H. & S.Fujimoto 1973. *Hydraulic Model Studies of Cristal Dam Spillway and Outlet Works.* Colorado River Storage Project, Bureau of Reclamation, Denver, Colo., Report no.REC-ERC-73-22.

Castro, P. 1979. An Investigation of Scale Effects in Vortex Modelling by Model to Prototype Comparison. Dissertation, Mechanical Engng. Dept., Worcester, Mass.

Chang, E. 1976. *Review of Literature on Drain Vortices in Cylindrical Tanks.* BHRA Report TN1342, March 1976.

Chang, E. 1977. *Review of Literature on the Formation and Modelling of Vortices in Rectangular Pump Sumps.* BHRA Report TN1414, June 1977.

Chang, E. 1979a. *Scaling Laws for Air-entraining Vortices.* BHRA Report RR1519, January 1979.

Chang, E. 1979b. *Experimental Data on the Hydraulic Design of Intakes and Rectangular Pump Sumps.* BHRA Report RR1518, January 1979.

Chang, E. 1980a. Application of a consistent vortex scaling law. *Proc. ASCE Hydraulics Specialty Conf., Chicago, August 1980.*

Chang, E. 1980b. An investigation into vortex formation in rectangular channels. In W.L.Swift et al. (ed.), *Vortex Flows, Proc. ASME Winter Annual Meeting.* Fluids Engineering Div., Chicago.

Chang, E. 1982. Model-Prototype Comparison of Free Surface Vortices; Discussion of Hecker (1981). ASCE. *J. Hydraul. Div.* 108, HY11.

Chang, E. & M.J.Prosser 1978. Intake design to Prevent Vortex Formation. *Proc. IAHR-ASME-ASCE Joint Symposium on Design and Operation of Fluid Machinery, CSU Fort Collins, June 1978*, vols I and III.
Chen, Y.N. 1979. From Bath Tub Vortex to Pump-Intake Vortex. Axial Flow-Induced Instability as their Formation Mechanism. *Schweizer Ingenieur und Architekt* 42:845-852.
Comninellis, E. 1973. Ludington Pumped Storage Project. ASCE. *J. Power Div.* 99, PO 1.
Daggett, L.L. & G.H.Keulegan 1974. Similitude in free-surface vortex formations. ASCE. *J. Hydraul. Div.* 100, HY11.
Dallwig, H.J., R.C.M.Schröder & E.Zäschke 1979. Wirbelbildung an Entnahmebauwerken. *Wasser und Boden* 1979(4).
Denny, D.F. 1956. An Experimental Study of Air-Entraining Vortices in Pump Sumps. *Journal of the Institution of Mechanical Engineers* 170(2). London, UK.
Denny, D.F. & G.H.J.Young 1957. The Prevention of Vortices and Swirl at Intakes. *IAHR Congress Lissabon*, Paper C1.
De Siervi, F., H.C.Viguier, E.M.Greitzer & C.S.Tan 1982. Mechanisms of inlet-vortex formation. *J. Fluid. Mech.* 124:173-207.
Desserich, M.A. 1980. Hochwasserentlastungsstollen Wirzenbach Luzern. *Schweizer Ingenieur und Architekt* 98(35).
Dexter, R.B. & E.R.Zeigler 1978. Penstock Intake Vortex and Related Turbine Operation Model Studies. *Proc. IAHR-ASME-ASCE Joint Symposium on Design and Operation of Fluid Machinery, CSU Fort Collins, June 1978*, vols I and III.
Dhillon, G.S. 1980. *Vortex Formation at Pipe Intakes and its Prediction.* Central Board of Irrigation and Power, New Delhi.
Dhillon, G.S., V.S.Sakhuja & T.C.Paul 1981. Modelling criteria for vortex formation at pipe intakes. *IAHR Congress New Delhi*, Paper D(b)-7.
Dicmas, J.L. 1967. Development of an Optimum Sump Design for Propeller and Mixed-Flow Pumps. ASME Paper 67-FE-26, presented at the Fluids Engineering Conference, Chicago, Illinois.
Dicmas, J.L. 1978. Effect of Intake Structure Modifications on the Hydraulic Performance of a Mixed Flow Pump. *Proc. IAHR-ASME-ASCE Joint Symposium on Design and Operation of Fluid Machinery, CSU Fort Collins, June 1978*, vols I and III.
Donnelly, C.A., G.G.Hebaus & F.W.Blaisdell 1974. *Hydraulics of closed conduit spillways, part XII: the two-way drop inlet with a flat bottom.* U.S. Dept. of Agric., Agric. Research Service, Minneapolis, Minn., Sept. 1974, ARS-NC-14.
Drioli, C. 1969a. Installazioni con Pozzo di Scarice a Vortice. *L'Energia Elettrica* 46(2).
Drioli, C. 1969b. Esperience su Installationi con Pozzo di Scarico a Vortice. *L'Energia Elettrica* 46(6).
Durgin, W.W. & G.E.Hecker 1978. The Modelling of Vortices at Intake Structures. *Proc IAHR-ASME-ASCE Joint Symposium on Design and Operation of Fluid Machinery, CSU Fort Collins, June 1978*, vols I and III.
Einstein, H.A. & H.L.Li 1955. Steady Vortex in a Real Fluid. *La Houille Blanche* 1955(4).
Elder, R.A., D.Hay, A.Alsaffar & O.Batum 1978. Vortex Model Studies for Emergency Core Cooling Intakes. *Proc. IAHR-ASME-ASCE Joint Symposium on Design and Operation of Fluid Machinery, CSU Fort Collins, June 1978*, vols I and III.
Farell, C. & A.R.Cuomo 1981. *Introduction to the Study of Vortex Dynamics and Similitude in Free Surface Flows.* Cenvênio Departamento de Aguas e Energia Elétrica e Escola Politécnica da Universidade de São Paulo.
Farell, C. & A.R.Cuomo 1982. Model-Prototype Comparison of Free Surface Vortices; Discussion of Hecker (1981), ASCE. *J. Hydraul. Div.* 108, HY11.
Florjancic, D. 1970. Influence of Gas and Air Admission on the Behavior of Single and Multi-Stage Pumps. Sulzer Research no.1970.
Fraser, W.H. 1953. Hydraulic Problems Encountered in Intake Structures of Vertical Wet-Pit Pumps and Methods Leading to their Solution. *Trans ASME* 75(4).

Gaspard, B., L.Netter, E.Becker & P.Filippi 1949. L'emissaire Sud d'Orly. *Travaux* 33(6).
Giesecke, J. & H.Brombach 1974. Strömungsmechanische Elemente im Wasserbau. *Wasser und Boden* 26(12).
Gordon, J.L. 1970. Vortices at Intakes. *Water Power* 1970(4).
Granger, R. 1966. Steady Three-Dimensional Vortex Flow. *J. Fluid Mech.* 25(3):557-576.
Gulliver, J.S. & A.J.Rindels 1985. Weak Vortices at Vertical Intakes. *Journal IAHR* 1985.
Gulliver, J.S., A.Rindels & K.C.Lindblom 1983. Guidelines for Intake Design Without free Surface Vortices. *Waterpower '83, Proc. Intern. Conf. Hydropower, Knoxville, TN.*, vol.III:1472-1483.
Gulliver, J.S., A.Rindels & K.C.Lindblom 1984. *Case Studies of Hydroplant Intake Design to Avoid Free-Surface Vortices.* St.Anthony Falls Hydraulic Laboratory, University of Minnesota, Project Report no.226.
Haindl, K. 1959. Contribution to Air-Entrainment by a Vortex. *IAHR Congress Montreal*, Paper D16.
Hattersley, R.T. 1965. Hydraulic Design of Pump Intakes. ASCE. *J. Hydraul. Div.* 107, HY2.
Hausser, R. & G.Galliere 1960. *Collectyeur D'Ecute Rainbault.* LaSalle Hydraulics Laboratory Report no.NC-229.
Hecker, G.E. 1981. Model-Prototype Comparison of Free Surface Vortices. ASCE. *J. Hydraul. Div.* HY10.
Hecker, G.E. 1984. Scale effects in Modelling Vortices. *Proceedings of the IAHR/DVWK Symposium on Scale Effects in Modelling Hydraulic Structures*, Paper 6.1. Technische Akademie Esslingen.
Hecker, G.E. & J.Larsen 1983. *Turbine Reaction to Free Surface Vortices.* Electric Power Research Institute, Palo Alto, California. Report no.EPRI EM-3017, Project 1199-8.
Heinemann, E. 1979. *Beitrag zur Vermeidung von Wirbelbildung vor Tauchwaenden.* Institut für Wasserbau und Wasserwirtschaft, TH Aachen, Aachen, Mitteilung no.27.
Heinemann, E. & G.Rouve 1977. Vortex Formation in Two-Phase Free Surface Flow. *IAHR Congress Baden-Baden*, Paper A57.
Herstrom, A.C., L.A.Polivka & R.Russell 1982. A Fourth Unit for the Mayfield Hydro Project. *Water Power* 1982(7).
Holtorff, G. 1964. La surface libre et les conditions de similitude du vortex. *La Houille Blanche* 1964(3).
Horeni, P. 1984. *Protection of Pump Sumps against Vortices.* Water Research Institute Praha, Práce a Studie 163.
Hughes, R.L. 1975. Similitude in Free-Surface Vortex Formations; Discussion of Daggett & Keulegan (1974). ASCE. *J. Hydraul. Div.* 101, HY9.
Humphreys, H.W., G.G.Sigurdson & H.J.Owen 1970. *Model Test Results of Circular, Square, and Rectangular Forms of Drop-Inlet Entrance to Closed-Conduit Spillways.* Illinois State Water Survey, Report of Investigation 65.
Hydraulic Institute Standards 1975. *Centrifugal, Rotary and Reciprocating Pumps.* 13th edition. Cleveland, Ohio.
Iversen, H.W. 1953. *Studies of submergence requirements of high specific speed pumps.* Univ. of Calif., Inst. of Eng. Research, Berkeley.
Jain, A.K. 1981. Physical Modelling of Vortices at Intakes. *IAHR Congress New Delhi*, Subject D(b)-16.
Jain, A.K., K.G.Ranga Raju & R.J.Garde 1978. Vortex Formation at Vertical Pipe Intakes. ASCE. *J. Hydraul. Div.* 104, HY10.
Jain, S.C. 1984. Tangential Vortex-Inlet. ASCE. *J. Hydraul. Engng.* 110(12).
Jain, S.C. & J.F.Kennedy 1983. *Vortex-Flow Drop Structures for the Milwaukee Metropolitan Sewerage District Inline Storage System.* Institute of Hydraulic Research, The University of Iowa, Iowa City, Iowa, July, IAHR report no.264.
Jain, S.C. & J.F.Kennedy 1984. Vortex-Flow Drop Structures. *Proceedings of the IAHR/DVWK Symposium on Scale Effects in Modelling Hydraulic Structures*, Paper 6.7. Technische Akademie Esslingen.

Jeanpierre, D. & A.Lachal 1966. Dissipation d'energie dans un puits a vortex. *La Houille Blanche* 1966(7).
Jevdjevich, V. & L.Levin 1953. Entrainment of Air in Flowing Water and Technical Problems Related with it. *Proc. Minnesota International Hydraulic Convention.* IAHR.
Johnson, P.L. 1972. *Hydraulic Model Studies of the Forebay Reservoir Inlet-Outlet Structure for Mt.Elbert Pumped-Storage Powerplant, Fryingpan-Arkansas Project, Colo., USA.* Bureau of Reclamation, Report Rec-Erc-72-5.
Kamel, M.Y.M. 1964. *The effect of swirl on the flow of liquids through bottom outlets.* ASME Paper 64-WA/FE-37.
Kenn, M.J. & K.J.Zanker 1967. Aspects of similarity for air-entraining water flows. *Nature* 213(5071):59-60.
Kleinschroth, A. 1972. *Strömungsvorgänge im Wirbelfallschacht.* Institut für Hydraulik und Gewässerkunde, TU München, Mitt. no.8.
Kleinschroth, A. & M.Wirth 1981. *Dimensionierung von Wirbelfallschächten unter Berücksichtigung verschiedener Zuflussbedingungen.* Institut für Hydraulik und Gewässerkunde, TU München, Mitt. no.36.
Knapp, F.H. 1960. *Ausfluss, Uberfall und Durchfluss im Wasserbau.* Braun, Karlsruhe.
Knauss, J. 1972a. *Wirbel an Einläufen zu Wasserkraftanlagen.* Versuchsanstalt für Wasserbau der Technischen Universität München, Bericht no.24.
Knauss, J. 1972b. Vortices and Swirling Flow at Low Head Intakes in Pumped Storage Schemes. *Un-ECE Symposium on Hydro-electric Pumped Storage Schemes, Athens,* Paper C2.
Knauss, J. 1983. *Wirbelbildung an Einlaufbauwerken (Luft- und Dralleintrag).* DVWK-Schriften, Heft 63.
Kohlhase, S. 1973. *Uber den Vordrall an Pumpeneinläufen im Wirbelfeld.* Franzius-Institut, TU Hannover, Mitteilung no.38.
Kolf, R.C. & P.B.Zielinski 1959. The Vortex Chamber as an Automatic Flow Control Device. ASCE. *J. Hydraul. Div.* HY12.
Krause-Klein, Th. 1984. *Schwimmstoffrückhalt an festen Tauchwänden unter besonderer Berücksichtigung der Wirbelbildung.* Institut für Wasserbau und Wasserwirtschaft, TH Aachen, Mitteilungen no.51.
Kuznetsova, E.V. 1975. Vortex-Flow Shaft Spillways for High-Pressure Hydropower Schemes. *Gidrotekhnicheskoe Stroitel'stvo* 5 (translated for the American Society of Civil Engineers).
Larsen, J. & J.F.Noreika 1983. *Hydraulic Model Investigations of the Cooling Tower Circulating Water Pump Sump for the St.John's River Power Park.* Alden Research Laboratory, Worcester Polytechnic Institute, Holden, Mass., Report no.68-83.
Laushey, L.M. 1952. *Flow in Vertical Shafts.* Carnegie Inst. of Technology, Dept. of Civil Engineering, Pittsburgh, Pa., Research Report.
Laushey, L.M. 1953. Studies Show Pittsburg How to Drop Sewage 90 ft Vertically to Tunnel Interceptors. *Engineering News Record,* March 5.
Laushey, L.M. & F.T.Mavis 1953. Air Entrained by Water Flowing Down Vertical Shafts. *Proc. Minnesota Internat. Hyd. Conv.* IAHR.
Lee, H.L. 1975. Swirling Flow in Inlets. M.S. thesis in Mechanical Engineering, Worcester Polytechnic Institute, Worcester, MA.
Levi, E. 1971. Efficiency of Air-Entraining Vortex Formation at Water Intakes; Discussion of Quick (1970). ASCE. *J. Hydraul. Div.* HY3.
Levi, E. 1972. Experiments on Unstable Vortices. ASCE. *J. Engng. Mech. Div.* EM3.
Lewellen, W.S. 1962. A Solution for Three-Dimensional Vortex Flows with Strong Circulation. *J. Fluid Mech.* 14(3):420.
Linford, A. 1965. *The Application of Models to Hydraulic Engineering, Part 2: Air Entraining Vortices, Water and Water Engineering,* pp.105-110. London, UK, Mar.
Markland, E., J.McCorquodale & H.O.Anwar 1969. Effects of Viscosity on Vortex-Orifice Flow; Discussion of Zielinski & Villemonte (1968). ASCE. *J. Hydraul. Div.* HY1.

Markland, E. & J.A.Pope 1956. Experiments on a Small Pump Suction Well, with Particular Reference to Vortex Formations. *Inst. of Mechanical Engs.* 170(2):95-105.

Marris, A.W. 1967. Theory of the Bathtub Vortex, Journal of Applied Mechanics, Transactions of the ASME, March, pp.11-15.

Martins, R. 1984. Vortex Similitude in Hydraulic Models. *Proceedings of the IAHR/DVWK Symposium on Scale Effects in Modelling Hydraulic Structures*, Paper 6.9. Technische Akademie Esslingen.

Maynord, S.T. 1982. *Blue Waters Ditch Pumping Station East St.Louis, Illinois.* WES Vicksburg, Miss., Technical Report HL-82-13.

McCorquodale, J. 1968. Scale Effects in Swirling Flow. ASCE. *J. Hydraul. Div.* HY1.

Murakami, M., H.Suehiro, T.Isaji & J.Kajita 1969. Flow of Entrained Air in Centrifugal Pumps. *IAHR Congress Kioto*, Paper B8.

Nakato, T., J.F.Kennedy & L.Pruitt 1978. Model Investigation of the Lake Chicot Pumping Plant. *Proc. IAHR-ASME-ASCE Joint Symposium on Design and Operation of Fluid Machinery, CSU Fort Collins, June 1978*, vols I and III.

Nathan, W. 1981. Flow Investigations in Two Pumping Pits. ASCE. *J. Hydraul. Div.* HY1 (tech. notes).

Nelson, M.E. & H.J.Johnson 1959. Vortex Problems at Intakes for Lock Hydraulic Systems. *IAHR Congress Montreal*, Paper 12B.

Nystrom, J.B., M.Padmanabhan & G.E.Hecker 1982. Modelling flow characteristics of reactor sumps. ASCE. *J. Energy Div.* 108, EY3.

Odgaard, A.J. 1984. Fluid Properties and their Scale Effects in Froude-Scaled Hydraulic Models. *Proceedings of the IAHR/DVWK Symposium on Scale Effects in Modelling Hydraulic Structures*, Paper 6.10. Technische Akademie Esslingen.

Padmanabhan, M. 1978. *Selection and Scaling of Horizontal Gratings for Vortex Suppression.* Alden Research Laboratory, Worcester Polytechnic Institute, Holden, Mass., July 1978.

Padmanabhan, M. 1982a. *A Parametric Study of the Hydraulic Performance of Containment Emergency Sumps with Vertical Double-Outlets.* Alden Research Laboratory, Worcester Polytechnic Institute, Holden, Mass., Report no.47A-82.

Padmanabhan, M. 1982b. *Hydraulic Performance of Pump Suction Lines of Emergency Core Cooling Systems in Boiling Water Reactors*. Alden Research Laboratory, Worcester Polytechnic Institute, Holden, Mass., Report no.50-82.

Padmanabhan, M. 1984. Air Ingestion Due to Free Surface Vortices. ASCE. *J. Hydraul. Engng.* (tech. note).

Padmanabhan, M. & G.E.Hecker 1979. Flow Characteristics of Reactor Recirculation Sumps. *IAHR Congress Cagliari*, Paper C(d)-6.

Padmanabhan, M. & G.E.Hecker 1982. *Assessment of scale effects on vortexing, swirl and inlet losses in large scale sump models.* Alden Research Laboratory, Worcester, Polytechnic Institute, Holden, Mass., Report no.48-82/M398F.

Padmanabhan, M. & G.E.Hecker 1984. Scale Effects in Pump Sump Models. ASCE. *J. Hydraul. Engng.* 110, HY11.

Padmanabhan, M. & C.R.Janik 1980. Swirling Flow and Its Effect on Wall Pressure Drop Within Pipes. In W.L.Swift et al. (eds), *Vortex Flows, Proc. ASME Winter Annual Meeting.*

Papworth, M. 1975. *Dinorwich Pumped Storage Scheme, Lower Reservoir Model.* BHRA Fluid Eng., UK, Res. report RR 1283.

Patel, B.R. & P.W.Runstadler, Jr. 1978. Investigations into the Two-Phase Flow Behavior of Centrifugal Pumps, Polyphase Flow in Turbo-Machinery. *Proc. ASME Winter Annual Meeting, San Francisco, California.*

Paterson, I.S. & R.M.Noble 1982. The Right Approach. *Proceedings of the IAHR-Symposium on Operating Problems of Pump Stations and Power Plants.* Amsterdam.

Pennino, B.J. & G.E.Hecker 1979. A Synthesis of Model Data for Pumped Storage Intakes, Pump Turbine Schemes: Planning, Design, and Operation. *Proceedings of the Joint ASME/CSME Applied Mechanics, Fluids Engineering and Bioengineering Conference.*

Pennino, B.J. & J.Larsen 1982. *Experimental Evaluation of a Pump Station Expansion, Jefferson Parish Pumping Station No.2.* Alden Research Laboratory, Worcester Polytechnical Institute, Holden, Mass., Report no.98-82.
Pica, M. 1969. Scale Effects in Swirling Flow; Discussion by McCorquodale (1968). ASCE. *J. Hydraul. Div.* 95, HY1.
Pica, M. 1970. Scaricatori a Vortice. *L'Energia Elettrica* 47(4).
Plichon, J.N. & P.Serin 1959. Desaerateur a casserole resultats d'essais sur modele. *IAHR Congress Montreal.*
Polikovski, V.I. & R.T.Perelman 1956. The formation of vortices in a liquid with a free surface. *Gidrotekhnik* 5 (in Russian).
Prosser, M.J. 1977. *The Hydraulic Design of Pump Sumps and Intakes.* British Hydromechanics Research Association/Construction Industry Research and Information Association.
Quick, M.C. 1961. The Annular Hydraulic Jump. *Civil Engineering and Public Works Review* 56(662).
Quick, M.C. 1962. Scale Relationships between Geometrically Similar Free Spiral Vortices. *Civil Engineering and Public Works Review*, Sept./Oct. 1962.
Quick, M.C. 1970. Efficiency of Air-entraining Vortex Formation at Water Intakes. ASCE. *J. Hydraul. Div.* HY7.
Rahm, L. 1953. *Flow Problems with Respect to Intakes and Tunnels of Swedish Hydroelectric Power Plants.* Trans. Royal Inst. of Tech., Stockholm, no.71.
Reddy, Y.R. & J.A.Pickford 1972. Vortices at Intakes in Conventional Sumps. *Water Power* 1972(3).
Reddy, Y.R. & J.A.Pickford 1974. Vortex Suppression in Stilling Pond Overflow. ASCE. *J. Hydraul. Div.* HY11.
Rindels, A.J. & J.S.Gulliver 1983. *An Experimental Study of Critical Submergence to Avoid Free-Surface Vortices at Vertical Intakes.* St.Anthony Falls Hydraulic Laboratory, Minneapolis, Minn., Report no.224.
Rohan, K. 1965. Velocity distribution in a free vortex. *Vodohospod. Cas.* 13(3):345-363 (in Czech).
Rohan, K. 1966. Conditions of similarity for a drain vortex. *Symposia CWPRS, Poona*, vol.2, pp.89-93.
Schulthess, D. 1984. Model Tests with Vortices — El Cajon Project (Honduras). *Proceedings of the IAHR/DVWK Symposium on Scale Effects in Modelling Hydraulic Structures*, Paper 6.5. Technische Akademie Esslingen.
Schulthess, D. 1985. Vortex Tower for the El Cajon spillway intake. *Water Power* 1985(7).
Singhal, M.K., J.Mohan & S.S.Tiagi 1979. Low Level Operational Problems of Outlet Conduits. *ICOLD Congress New Delhi*, Q.50, R.48.
Slissky, S.M., E.V.Kuznetsova & T.Kh.Akhmedov 1979. Multi-Stored Vortex Shaft Outlets. *ICOLD Congress, New Delhi.*
Springer, E.K. & F.M.Patterson 1969. *Experimental Investigation of Critical Submergence for Vortexing in a Vertical Cylindrical Tank.* ASME, Paper 69-FE-49, June 1969.
Spurr, K.J.W. 1980. Modelling in design of pumping pits; Discussion of Tullis (1979). ASCE. *J. Hydraul. Div.* 106, HY10.
Spurr, K.J.W. & F.J.Boucherat 1976. The Application of Vorticity Concepts to Three Irrigation Intake Designs. *Proc. Inst Civ. Eng., London, UK*, Part 2, June 1976, pp.311-330.
Stevens, J.L. & R.C.Kolf 1957. Vortex Flow Through Horizontal Orifices. ASCE. *J. Sanit. Engng. Div.* SA 6.
Stevens, J.L. & R.C.Kolf 1959. *Vortex flow through horizontal orifices.* Trans. ASCE, 124, Paper 3004, pp.871-883.
Sweeney, C.E., R.A.Elder & D.Hay 1982. Pump Sump Design Experience: Summary. ASCE. *J. Hydraul. Div.* 108.
Toyokura, T. & S.Akaike 1970. Vortex phenomena in a water tank. *Bull. JSME* 13(57):373-381.
Toyokura, T. & K.Kamemoto 1984. Introduction of the JSME Standard to Establish a Set of Basic Rules for Model Testing of Pump Sump Performance. *Proceedings of the IAHR/DVWK*

Symposium on Scale Effects in Modelling Hydraulic Structures, Paper 6.3. Technische Akademie Essingen.

Tullis, J.P. 1979. Modelling in Design of Pumping Pits. ASCE. *J. Hydraul. Div.* 105, HY9.

TVA Report 1973. *Tenessee Valley Auth. Report.* Div. Wat. Control Planning, Eng. Lab., USA, Annual Report 1971-2, Apr. 1973.

VAW Reports. Various Hydraulic Model Test Reports. Laboratory of Hydraulics, Hydrology and Glaciology of the Swiss Federal Institute of Technology, Zürich.

Venkataramana, S., P.J.Dinakar & B.H.Rajan 1984. On the Conflicting Modelling Criteria, Causative Scale Effects and the (in) Compatibility of Scale Modelling of Vortices — A Case Study. *Proceedings of the IAHR/DVWK Symposium on Scale Effects in Modelling Hydraulic Structures*, Paper 6.8. Technische Akademie Esslingen.

Venkatraman, C.P., J.K.Shanker Rao, V.Ramanathan & A.R.Chaven 1981. Experimental Investigations of Scale Effects in Vortex Formation at Intakes. *IAHR Congress New Delhi*, Paper D(b)-8.

Viparelli, M. 1950. Su un particolare tipo d'imbocco e sull efflusso con vortice. *L'Energia Elettrica* 27(10).

Viparelli, M. 1961. Les courants d'air et d'eau dans les puits verticaux. *La Houille Blanche* 1961(6).

Vischer, D. 1976. *Hydraulische Untersuchungen für die Abwassertechnik.* Tagungsbericht des Verbandes Schweizerischer Abwasserfachleute Bern, No.131/1.

Wang, J.D. & J.J.Connor 1975. *Mathematical Simulation of Near Coastal Circulation*, R.M.Parsons Laboratory for Water Resources and Hydrodynamics, MIT, Cambridge, MA., Technical report 200.

Weller, J.A. 1975. Similitude in Free-Surface Vortex Formations; Discussion of Daggett & Keulegan (1974). ASCE. *J. Hydraul. Div.* 101, HY11.

Weske, J.R. 1958. *On the Origin and Mechanism of Vortex Motion at the Inlet of Intakes Placed Near a Flat Surface.* University of Maryland, Tech. note BN 152, AFOSR-TN-58' 863.

Whillock, A.T. & M.F.C.Thorn 1973. *Air Entrainment in Dropshafts.* Construction Industry Research and Information Assoc., Technical note 48.

White, D.K. & J.Larsen 1979. *Hydraulic Model Investigation of the Existing Dilution Water Pump Intake, Northport Power Station.* Alden Research Laboratory, Worcester Polytechnic Institute, Holden, Mass., Report no.22-79.

White, D.K. & J.Larsen 1983. *Hydraulic Investigation of Flow Conditions in a Vertical Pump Intake Structure.* Alden Research Laboratory, Worcester Polytechnic Institute, Holden, Mass., Report no.87-83.

Whittles, K.G. 1982. Experimental Studies of Surface Vortices at a Vertical Intake. M.S. thesis, Worcester Polytechnic Institute, Worcester, MA.

Wijdieks, J. 1965. *Hydraulic Aspects of the Design of Pump Installations.* Delft Hydraulic Laboratory.

Wijdieks, J. 1984. Practice and Criteria of Pump Sump Model Investigation. *Proceedings of the IAHR/DVWK Symposium on Scale Effects in Modelling Hydraulic Structures*, Paper 6.4. Technische Akademie Esslingen.

Woodford, D.J. 1964. Prediction of the length of the air core vortices of weak circulation. *La Houille Blanche* 1964(8).

Yildirim, N. & S.C.Jain 1979. Effect of a Surface Layer on Free Surface Vortex. *IAHR Congress Cagliari*, Paper C(c)-24.

Yildirim, N. & S.C.Jain 1981. Surface Tension Effect on Profile of a Free Vortex. ASCE. *J. Hydraul. Div.* HY1 (tech. notes).

Zajdlik, M. 1977. New Checking Mode of Model Parameters for Vortex Formation in Pump Tanks. *IAHR Congress Baden-Baden*, Paper D(d)-49.

Zarate, F.J. & J.C.de Lio 1979. Aspects of Hydraulic Design for Power Plant Intake. *IAHR Congress Cagliari*, Paper C(a)-15.

Zielinski, P.B. & J.R.Villemonte 1968. Effects of Viscosity on Vortex-Orifice-Flow. ASCE. *J. Hydraul. Div.* 94, HY3.

Subject index